FUTURE YET TO COME

FUTURE YET TO COME

Sociotechnical Imaginaries in Modern Korea

Edited by Sonja M. Kim and Robert Ji-Song Ku

University of Hawai'i Press
Honolulu

Paperback edition 2022

Printed in the United States of America

Library of Congress Cataloging-in-Publication Data

Names: Kim, Sonja M., editor. | Ku, Robert Ji-Song, editor.
Title: Future yet to come : sociotechnical imaginaries in modern Korea / edited by Sonja M. Kim and Robert Ji-Song Ku.
Description: Honolulu : University of Hawaiʻi Press, 2021. | Includes bibliographical references and index.
Identifiers: LCCN 2021030659 | ISBN 9780824889197 (cloth) | ISBN 9780824889609 (pdf) | ISBN 9780824889616 (epub) | ISBN 9780824889623 (kindle edition)
Subjects: LCSH: Science—Social aspects—Korea (South)—History. | Technology—Social aspects—Korea (South)—History. | Social medicine—Korea (South)—History.
Classification: LCC Q175.52.K6 F88 2021 | DDC 306.4/5095195—dc23
LC record available at https://lccn.loc.gov/2021030659

ISBN: 9780824889630 (paperback)

Cover design: Aaron Lee

University of Hawaiʻi Press books are printed on acid-free paper and meet the guidelines for permanence and durability of the Council on Library Resources.

Contents

Acknowledgments

This book would not have been possible without the generous contribution of a great number of people. We would like to specifically express our gratitude to the following friends and colleagues who have provided invaluable support over the years as this book came together: John Cheng, Sungdai Cho, Fa-ti Fan, Hoe-Kyeung Kim, Immanuel Kim, Sang-Hyun Kim, Tae-Ho Kim, Katherine Martineau, Aurélien Laroulandie, Yoonkyung Lee, Giovanna Montenegro, Young-Gyung Paik, Yunjae Park, Michael Pettid, Soyoung Suh, and Yi Wang.

We are also grateful to our editor at the University of Hawai'i Press, Stephanie Chun, whose steadfast interest and support sustained this project. We also thank the two anonymous readers of the manuscript. This book is undoubtedly stronger due to their critical suggestions, and each chapter benefited greatly from their generous and constructive feedback. We thank the contributors for their expertise, acumen, and patience. This book would not exist without them.

This book was supported by Binghamton University's Center for Korean Studies, as well as the Core University Program for Korean Studies through the Ministry of Education of the Republic of Korea and Korean Studies Promotion Service of the Academy of Korean Studies (AKS-2011-BAA-2103).

Chapter 4, "The Suicidal Person: The Medicalization and Gendering of Suicide in Colonial Korea," by Theodore Jun Yoo, is a significantly revised version of "Madness as a Social Epidemic," which is chapter 4 of his book *It's Madness: The Politics of Mental Health in Colonial Korea* (2016).

Lastly, Sonja would like to dedicate this book to Jae and Milena.

Robert, as always, dedicates this book to Nancy, Oliver, and Eliot.

Notes on the Text

Throughout the volume, we employ the McCune-Reischauer system to romanize Korean terms and names. Chinese is rendered in Pinyin and Japanese in Hepburn, respectively. Exceptions are made if they have alternate spellings that are well established in the English language (e.g., Seoul instead of Sŏ'ŭl). Direct quotations and citations are transliterated as they were written in the original. East Asian names follow the standard sequence of family name first without using a comma. Given names are listed first for authors and artists who publish and exhibit in English. Translations are of the authors unless otherwise specified.

FUTURE YET TO COME

Introduction

Sonja M. Kim

Shortly after releasing the highly marketed premium Galaxy Note 7 smartphone in the late summer of 2016, Samsung Electronics recalled the devices and ceased production in response to concerns over fire risks posed by its lithium-ion battery. Customers replaced their phones and filed claims for damage and "economic injuries."[1] Airlines checked that the devices did not board certain flights.[2] Samsung's profits and its reputation as innovator and major player in consumer, digital, media, and information technologies were declared at risk. South Koreans watching the debacle unfurl held mixed feelings. On one hand, it was humiliating. Samsung's lion presence in the country's sociopolitical economy and global market earned South Korea the moniker the Republic of Samsung.[3] As one oppositional leader noted, "Because people take pride in Samsung as a brand representing South Korea, it is their trouble too."[4]

Not everyone agrees. A South Korean observer was reported to note, "The saying that Samsung's good and bad luck is our country's good and bad luck is propaganda manufactured by Samsung and media and politicians beholden to it."[5] To critics, the recall presented an opportunity to institute measures for corporate transparency, quality control, customer service, and better management for a stronger and perhaps more democratic future of the country.[6] The corruption trial the following year of Lee Jae-yong, the third-generation heir apparent to the Samsung Group and vice chairman of Samsung Electronics, further fanned broader public concerns of entrenched collusion and shady dealings between government and the private sector. Lee was convicted of bribing impeached President Park Geun-hye, but an appeal court's reduction and later suspension of his sentence laid bare struggles in South Korea's political system to disentangle conglomerates from nexuses of state power.[7]

Underneath explanations that attribute such difficulties to anxieties that any destabilization of Samsung would threaten the country's competitiveness in the world economy lies a longer history of state-led strategies. Forged in the authoritarian and developmentalist thrust of the decades following the Korean War (1950–1953), science and technology became associated with national strength and identity, to be realized by tactical selection of certain corporations and industries, such as shipbuilding, automobiles, and electronics. South Korea's pursuits in science, technology, and medicine (STM) are popularly understood as key elements in the country's transformation over the past half century from a war-torn, diseased, impoverished, and former Japanese colony to a global economic giant.[8] Korean scholars analyze in particular the intimate relationship since the mid-1960s between the South Korean state and private sector in the promotion of specific STM projects for economic developmental and national sovereignty or nation-building purposes.[9] The slogan *kwahak ipkuk* (nation-building through science) during Park Chung-hee's administration (1963–1979) is a form of what Hiromi Mizuno notes as "scientific nationalism," the mobilization of science or the scientific for the nation (or, as in her study, the empire) and in turn mobilization of the nation for science.[10]

The correlation articulated between scientific and technological achievements and the country's future relates to what Sheila Jasanoff and Sang-Hyun Kim call "sociotechnical imaginaries." Drawing from a rich literature in the social sciences on the imaginary, Jasanoff and Kim define sociotechnical imaginaries as the "collectively held, institutionally stabilized, and publicly performed visions of desirable futures, animated by shared understandings of forms of social life and social order attainable through, and supportive of, advances in science and technology."[11] Projections of desired (and sometimes prescribed) visions for the community, society, or nation generate and are generated in part by the sciences and their associated practices, medicine and technology.

Residing "in the reservoir of norms and discourses, metaphors and cultural meanings," sociotechnical imaginaries often relate but are not restricted to state power.[12] Jasanoff and Kim note that imaginaries are not necessarily static, singular, or complete; any given society may have multiple imaginaries. They may be articulated by nation-states as well as by other organized groups, such as civic associations and corporations, and even individuals. They become dominant and perform when buttressed by institutions of power, including courts and the media. They may then inform policies that stimulate or stymie the scientific and technological practices presented to realize those very visions. They also reflect and shape popular attitudes toward techno-science and those who practice it. At its heart, sociotechnical imaginaries "encode not only visions of what is attainable through science and technology but also of how life ought, or ought not, to be lived; in this respect they express a society's shared understandings of good and evil."[13] Accordingly, they allow for productive examinations of power and morality in the social and political orders in which they are embedded.

It is to these perspectives that this volume attests. Although the methodologies and subject matters taken span the diverse disciplines of the contributors, the essays together situate an understanding of science-related engagements as key constituents of social life and as cultural practices that, as Fa-ti Fan explains, are best understood in the particular contexts within which they unfold.[14] For example, as Mizuno reminds us, what nation or nationalism means varies because they do not have "an already existing or defined set of ideas, vocabularies, and programs."[15] She urges that science be comprehended as actions in a constantly changing material world rather than as an absolute or a single coherent entity.

If imaginaries structure and co-constitute social and cultural practices and thus STM activities, then imaginaries may produce (or are believed able to produce) material effects. What makes thinking critically about the sciences difficult, however, is the epistemological legitimacy they wield, often treated as objective and credible and as having universal applicability. Daniel Kleinman describes the array of ideas, concepts, categories, beliefs, and discourses that grant such cognitive authority to science and technology as "scientism" and "technological progressivism."[16] Korean scholars such as Sang-yong Song also note the primacy of and admiration for notions of science as progressive in Korea.[17]

In this framework, identifying STM as practices in action relates to agency. Postcolonial and science, technology, and society studies (STS) scholars problematize premising the STM techniques and values prevalent in historical narratives of Western Europe and North America as useful, natural, progressive, desired, or even universally significant. Doing so peripheralizes non-Western societies when they do not yield the same measures or pursue the same activities. Moreover, applying these standards potentially casts those societies as miracles, late, or failures, either having or not having latent potential.[18] Such narratives flatten histories of STM in non-Western societies by narrowing the purview of intellectual inquiry to reception of knowledge, goods, and practices that radiate from Western Europe and North America. They overlook the "creative process of appropriation, translation, and innovation."[19] Moreover, they neglect to consider other possibilities or validate intentional rejection of certain knowledge and methods. For example, Francesca Bray argues that Ming China (1368–1644) chose to place its resources elsewhere rather than navigate transoceanic voyages despite being able to do so. Bray also urges an expanded reading of technology as culture that includes symbolic and ideological work, a perspective that is lost when focus is placed on economic and material efficiency as defining features of technology.[20]

For scholars of Korea, the task of "de-peripheralizing" becomes more urgent given that the history of STM in Korea is further overwhelmed by the specter of China before the twentieth century and by Japan and the United States afterward. Although teleological models of modernization loom large in historical narratives of Korea, that many of the theories and practices related to STM in pre-industrial Korea came from, or were informed by Korea's relations with, China is not in doubt. Kim Yung Sik argues that this oriented early histories of Korean science toward an emphasis on the creativity and originality of Korea's scientific achievements.[21] They favored making prominent certain techniques and artifacts, such as printing technologies, the rain gauge, ceramics, *kusŏn* (the iron-clad armored warship usually known as the turtle ship or *kŏbuksŏn* developed in the fifteenth century), Chŏmsŏngdae (a seventh-century architectural structure presumed to have been an astronomical observatory), and even Hangŭl (the vernacular phonetic script), over ideas and institutions. We also witness assertions of Korean innovation in the realm of medicine.[22]

The showcasing of testaments of Korea's scientific achievements is well exemplified in South Korea's National Museum exhibits, elementary school textbooks, and Cultural Heritage Administration, knitting them to sentiments of national pride and reputation. Even the academic organization Korean History of Science Society (Hanguk kwahaksa hakhoe) in its earlier years devoted its associational conferences to highlight Korean innovation through artifacts and techniques: Korean movable metal type for printing in 1972, Chŏmsŏngdae in 1973, late Chosŏn theories of earth's rotation in 1975, and *kusŏn* 1976.[23] In the early 1980s, the society published a report on what it termed "scientific cultural properties" (*kwahak munhwajae*), further stimulating investigation of historical relics and artifacts, some of which the state later designated as national cultural properties.

Such demonstrations, however, may potentially produce anachronistic conceptualizations or categorizations of Korea's intellectual and material practices. Kim Yung Sik sympathizes with Nathan Sivin's move away from approaches in historical study of Chinese sciences framed by Joseph Needham's why-not question or what Sivin calls the "Scientific Revolution Problem."[24] The privileging of the construction of Western science as natural, universal, and objective closes inquiries into the arrangements or contexts in which natural studies operated. Yet Kim does not advocate skirting the big questions altogether so long as scholars are cognizant of problematic assumptions that gird unequal comparisons and yield hasty conclusions.[25] Applying this to Korea, Kim and other Korean STS scholars have found it productive to investigate the content as well as the contexts in which science engaged. Using knowledge and material transmissions from China, Japan, or the West as examples, questions that scholars may investigate include what the channels of transmission were and what meanings we assign them, who among Koreans were interested and why, and which transmissions did they prefer, reject, master.[26] Moreover, transmissions are multidirectional and never-ending, their topographies continually changing.

This opens the field to other interventions such as textual analysis and sociological, anthropological, and cultural methods that assess STM within their social and intellectual contexts. The positioning of STM as legitimate subjects for research, subjects that had been relatively marginalized in Korean historical scholarship, much less global histories of STM, formalized STS as a growing academic field in Korea.[27] This facilitates cross-fertilization between disciplines and raises new questions in Korean studies, such as the relationship between STM and Confucian statecraft, intellectual trends, imperialist power configurations, international geopolitics, and the state.[28] It also expands our understanding of transfers (not only from the West but also within Asia), regional and global circulations, translations, and accommodations "across the networks of science" in a history of global connections.[29] If STS is inherently a postcolonial project to "decolonize science and technology studies" and interrogate the global as Warwick Anderson suggests, learning about local expressions of STM can tell us much about today's world and add to the heterogeneity of STS in general.[30] That many STS scholars participate actively in social movements does not surprise given the tendency in STS to critically examine STM and their political dimensions.

Future Yet to Come surveys diverse technoscientific activities and their intersections with imaginaries in Korea's long twentieth century. If we step back and explore how contingent STM activities are on their social, cultural, political, global, and historical contexts, we are better able to decipher the processes that enable them to gain the currency they do for Koreans and their communities over time and space. We see how they mark certain values or indicators, such as neoliberalism, democracy, or moral order. What sets this volume apart from other studies is its broader historical perspective and multidisciplinary approach that pays special attention to public health, life sciences, and the media arts.

Jasanoff and Kim acknowledge history's indeterminacy, human agency, and the mutually productive processes of STM and the social, yet they premise sociotechnical imaginaries on modernist projects that are oriented toward the future in nature. In Korean STS, much of the focus has been on the period after the Korean War and on large-scale state projects such as nuclear power.[31] This volume encourages a longer arc of history to highlight dialectical configurations of time. Anxieties evoked or desires that promote pursuits of a better future engage with the past to act in the present; modern imaginaries, for example, may draw from older ones, allowing the application of the concept to a pre-industrial era. Tracing imaginaries across the vicissitudes of Korea's tumultuous past reminds us of their history and makes visible shifts in their nature and shape over time.

Although some of the chapters do not articulate or center their analysis on imaginaries or technologies, together they use imaginaries as a heuristic to examine the contexts within which STM endeavors form. They assess the implications that shared visions pose for STM projects, society in general, and individuals in their everyday lives. Contributors acknowledge the close integration between STM activities and the state but also recognize multiple actors, imaginaries beyond nation-building centered ones, and desires other than economic growth intersecting with STM pursuits. The sciences, their processes and outcomes, are in many ways central to Korea's social and cultural imaginaries. They channel scripts for what it means to be human, how to act in one's body, in relation to others, to the material world, and in social practices. The volume thus features health and life sciences to examine how STM may enable different modes of the self and value (or not value) certain forms of life.

These discussions provide valuable insights on contemporary South Koreans' health experiences. Although the essays were written before the outbreak of the pandemic, they allow for comparative analysis such as the country's early handling of the 2019 coronavirus disease (COVID-19). Media accounts attribute South Korea's relative success in spring of 2020 to factors such as immediate widespread testing, comprehensive contact tracing, and strict social isolation or distancing.[32] Whether they are replicable, feasible, or would produce similar results in other contexts, such as the United States, remains uncertain. The speed and breadth with which South Koreans adopted their measures were enabled by close relations between the state and corporations, the national health-care system, historical experiences with large-scale public health campaigns that elicited the public's cooperation, and neoliberal promotions of the self, tied to imaginaries addressed in this volume.

Additionally, the realm of visual culture provides rich areas in which to examine Korean sociotechnical imaginaries in unexpected ways. The creative productions of individual artists or artist collectives may not qualify as imaginaries according to social theory unless they become communally adopted. Taking a broader perspective of technology as ways of making and doing in producing material or immaterial social value as Francesca Bray suggests, however, intimates the performance of artists as technicians and their aesthetic practices as technologies. Not only is their art enabled by developments in optics, chemistry, and telecommunication technologies that have become available, but also artists use their techniques in challenging or offering competing social imaginaries or moral visions for Korean communities in particular historical moments. They evoke various sensibilities and forms of "prosthetics" in the human experience, such as vision and memory, to remind selves and others of communal values, inspire action, or question. And when human interaction with technology becomes the subject matter such as in science fiction, art becomes "a repository of sociotechnical imaginaries."[33]

This book is organized thematically in three sections, with the chapters in each section arranged roughly chronologically. These distinctions, however, are not absolute, and overlap among chapters attests to the shared perspective among contributors. STM pursuits as cultural practices do not exist independently of but instead engage productively with the social, political, and cultural contexts in which they are formed and used. Incorporation of new technologies are situated amid global interactions and patterns. Scientific achievements are not finished products but instead reflect the fluid nature of knowledge "in motion." STM then become critical sites in illuminating societal transformations that take place, revealing as much about understandings of nature and the self as they do about Korean power relations, anxieties, desires, and values. The cases presented here serve as points of comparisons with other countries, for example, by reconsidering "benevolence" and "aid" with forms of violence embedded in their choices and preferences and reclaiming philosophy and ethics in Korean STM studies.

Our goal is to add insight by tracing operations of Korea's social imaginaries over time as they intersect with STM pursuits to live better tomorrow starting today. The title of the volume, *Future Yet to Come*, is inspired by Jieun Lee's chapter and alludes not so much to a teleological orientation toward the future as to the different configurations and dialectical relations of time experienced by Koreans where the past, present, and future overlap and inform one another along the transformative nature of Korea since the Chosŏn period (1392–1910). South Korea is the primary focus, but the volume explores "other" Koreas—for example, Chosŏn Korea, colonial Korea, and Korean diaspora—to provide context, juxtaposition, and contrast.[34] The challenge is to offer Korea as a "locus of enunciation" without parochializing or essentializing it.[35]

The volume neither answers definitively how practices in STM unfolded in Korea nor suggests policies. Nor does it examine the full range of STM activities with which Koreans participate. Instead, it offers snapshots of embodied experiences engendered by STM engagements, without placing the roles played by Japan in its colonization of Korea (1910–1945) or the United States or international aid organizations in the latter half of the century at center. Positioning historical and STS-oriented chapters with those in the creative arts reflects recent moves that examine intersections between science and literature.[36] Here, they allow us to observe imaginaries framed by the state at different historical moments in the first section, their operations in health and medical projects in the second, and their aesthetic expressions in the third. Together, they demonstrate the resilience of imaginaries in relation to the sciences.

Part I, "Recollecting Sociotechnical Imaginaries," lays out the kinds of imaginaries exercised by state power that informed and legitimated larger-scale STM projects from the Chosŏn dynasty to the present. Don

Baker opens the volume, noting that science at its basis is the knowledge used to comprehend the world and should not be restricted by contemporary epistemologies of science. Chosŏn sciences, though akin to natural philosophy and inseparable from the metaphysics of Confucian thought, reflect the search for and desire to manifest "patterns of appropriate interactions" humans are to have in the material world. In this way, rulers and thinkers conceived a natural world that was as moral as it was material. Instrumental to governance, sciences as organized by the state were moral endeavors and took shape in ways to meet Chosŏn ends in a specific Chosŏn environment.

Chosŏn concepts of nature would be severely challenged around the turn of the twentieth century, but, as Sonja Kim attests in chapter 2, moral imperatives that framed medicine as a virtuous art and informed social imaginaries during the Chosŏn period did not disappear completely. Medical practices took shape in new institutional settings with different intellectual authority, but moral imperatives were rearticulated in popular discourse and in relation to the political order between the Chosŏn and Japanese colonial periods. In chapter 3, Inkyu Kang asks how South Korea's worst scientific fraud and the public's quick forgiveness of Hwang Woo-suk's fabricated claims to successfully clone human embryos arose under a supposed liberal political administration in the twentieth-first century. The answer lay in part with the developmentalist legacy of South Korea's technological utopianism that identified science and technology as a way out of destitution and misery and toward national greatness. But another critical element, Kang argues, includes the deregulation and other changes in South Korea's educational infrastructure that laid the grounds on which academic misconduct was made possible and perhaps inadvertently encouraged. Kang's chapter demonstrates the resilience of sociotechnical imaginaries in changed political economies.

Part II, "Restoring Minds and Bodies," examines specific case studies of imaginaries in action. Here, the lens shifts to the valuing of life in medical sciences and public health. Bodily interventions in the realm of psychiatry, infectious disease control, rehabilitative medicine, and bio-insurance are part and parcel of projects to mobilize bodies for specific political and personal agendas during the colonial period, enmeshed in the geopolitics of the Cold War and ensuing international arrangements, South Korean state's developmentalist desires, and market speculations of the current era.

Moreover, the material goods produced, and bodily practices exhorted, help create and foster new forms of the self.

In chapter 4, Theodore Jun Yoo's analysis of competing discourses on suicide in colonial Korea reveals how language in the modernist fields of medicine and psychiatry gave rise to novel interpretations of suicide as pathological and popularized the image of the "suicidal person." Medical language, however, competed with sociological explanations, and social anxieties about the mentally ill as threats to the self and public were supported by the legal force and power buttressed by scientific knowledge. Jane Kim argues in chapter 5 that long-term goals to promote social and political stability inadvertently produced measures that countered accepted medical knowledge in 1950s South Korea. She demonstrates this with a study of public health campaigns against Hansen's disease (leprosy) as pursued by South Korean health authorities who were informed by the World Health Organization's recommendations. It is the afflicted patients who suffer, thwarted by the geopolitical goals of social and political stability, which were taken more seriously than therapeutic solutions.

In chapter 6, John DiMoia tackles the literal project of restoring bodies in tracing the development of rehabilitative medicine from its roots during the Korean War to its later iteration in child welfare and occupational therapy. This case study displays biomedicine as a critical part of postwar Korea-US relations. But, as DiMoia demonstrates, by participating in an international economy of medical exchange, Korean actors extended the scope of rehabilitation and applied the notion of disability to broader social categories of persons. Prostheses and medical assistance served to rehabilitate the "broken" bodies of war veterans and children into strong able-bodied citizens to celebrate South Korea's economic and political recovery in the industrial workplace. Decades later, prostheses in the form of stem cells, offer individuals "insurance" against a possible future of illness. In chapter 7, Jieun Lee argues that the speculative market for stem cell banking and what she calls bio-insurance creates new configurations of life, time, and money. Language used in advertising these services summon social anxieties about aging. Bio-insurance serves less to manage population than to normalize individuals as "*capable* citizens," who are interpellated as "not-yet-ill" subjects.

Part III, "Prosthetic Arts," borrowing the concept of prosthesis from the previous section and Steve Choe's chapter in this section, shifts gear to aesthetic expressions of imaginaries in the creative arts that function to interrogate and inspire action, utilizing modes facilitated by technological developments in the media arts. Technological and scientific developments, although important to industrialization and manufacturing processes, also produce different media that become available for artistic expression and novel ways of being human as we live with, use, and experience new technologies. They shape ways we perceive and relate with one another and how we remember history. In chapter 8, Hyeri Oh traces techniques enabled in photography by the fields of optics and chemistry to spark a new genre or art movement of realism in 1950s South Korea. Presented as the "scientific reproduction of reality," photography reflected a commitment to the social and historical conditions of post–Korean War South Korea in hopes of stimulating the sensibility of humanism among its viewers, and as a form of prosthetics. Communication technologies such as the television, according to Steve Choe in chapter 9, are another example of prosthetics. They mediate the way we imagine our communities as well as the past, a point made by the video images of the 1995 installation *Electric Superhighway* by multimedia Korean American artist Nam June Paik. Video screens in the installation *Megatron/Matrix* project local images (that is, of South Korea) that are both cosmopolitan and global with their familiarity in our tertiary memory. On one hand, video images and screens become the media in artistic expression. On the other, the art itself humanizes our relationship to technology.

In chapter 10, the final one in this volume, Haerin Shin tackles this very question—what happens when technology itself becomes the genre and topic and not just the method of expression? Shin argues that the 2012 Korean film trilogy *Doomsday Book* makes an affective turn in South Korean science fiction cinema, setting it apart from other films in this genre. Whereas the plot in other films revolves around resolution of a paranormal situation (for example, river monster created from the excesses of human pollution or using a time machine to change a warped history), *Doomsday Book* presents new alterity or doom presented by science and technology (viral-induced zombies, alien world and destroyed earth, enlightened robots) as the condition in which film characters live with and cope. On the surface, the films serve as critique or warning to the excesses of science. In the end, though, they address what "it means to live and survive in an age of uncontrollable, unintended technological development" that affect the "body's capacity to act, engage, and connect" with one another. Shin's analysis serves not so much as a conclusion as a reminder of how deeply intertwined science, technology, and medicine are to not only our polities, corporation, and societies but also the very human condition.

Notes

1 A class-action lawsuit in New Jersey was filed by customers faced with local carrier chargers while returning and replacing their phones. See Christina Warren, "Samsung Hit with Class Action Lawsuit over Note 7 Debacle," *Gizomodo,* October 18, 2016, https://gizmodo.com/samsung-hit-with-class-action-lawsuit-over-note-7-1787927496. See also John Ribeiro, "Samsung faces lawsuit from Note 7 owners who couldn't use their phones," *PC World,* October 19, 2016, https://www.pcworld.com/article/3132829/samsung-faces-lawsuit-from-note7-owners-who-couldn-t-use-their-phones.html.

2 The US Federal Aviation Authority banned the device from domestic flights. Other airlines soon followed suit, among them All Nippon Airways, Japan Airlines, Asiana, Cathay Pacific, Air Berlin, Alitalia, and Singapore Airlines. See Samuel Gibbs, "Samsung Galaxy Note 7: Airlines across Asia, the US and Europe Ban Exploding Smartphone," *The Guardian,* October 17, 2016, https://www.theguardian.com/technology/2016/oct/17/samsung-galaxy-note-7-airlines-asia-us-europe-ban-exploding-smartphone.

3 Chico Harlan, "In S. Korea, the Republic of Samsung," *Washington Post,* December 9, 2012, https://www.washingtonpost.com/world/in-s-korea-the-republic-of-samsung/2012/12/09/71215420-3de1-11e2-bca3-aadc9b7e29c5_story.html. When and how the term "Republic of Samsung" was first coined is not clear. However, it appears in South Korean mass media as early as 2005 and has been picked up by foreign presses as well. The term refers in general to Samsung's significant presence in South Korean society (from consumer products to apartments, hospitals, and credit cards) and economy (accounting for roughly one-fifth of the country's GDP). It also alludes to South Korea's system of family-run conglomerates or chaebŏl, of which Samsung is representative, and the power conglomerates derive in part from their close ties with government and politics that have often been the focus of social critique and concern. For an earlier discussion of the moniker and divided opinions held in South Korea in regards to Samsung, see Chang Myŏnghŭi, "'Hanguk=Samsŏng konghwaguk e konggam' 48.8%," *Sisa chŏnŏl* [Sisa Journal], September 9, 2005, http://www.sisapress.com/news/articleView.html?idxno=102929. I thank Tae-ho Kim for directing me to this article.

4 Quoted in Sang-Hun Choe, "Galaxy Note 7 Recall Dismays South Korea, the 'Republic of Samsung,'" *New York Times,* October 22, 2016, https://www.nytimes.com/2016/10/23/world/asia/galaxy-note-7-recall-south

-korea-samsung.html. A South Korean passenger waiting to board a plane "said he felt humiliated, as if the non-Koreans in the airport lounge were looking for him," even though he does not own a Note 7.

5 Ibid.

6 Editorial, "'Kaellŏksi notŭ 7 sagŏn' ŭro pungijŏm e sŏn Samsŏng chŏnja" [Samsung's Galaxy Note 7 and the crisis of losing consumers' trust], *Hankyoreh,* October 11, 2016, http://www.hani.co.kr/arti/opinion/editorial/765196.html. See also "Yi Chaeyong puhoejang ŭi ŏkkae e wigi" [To save its brand, Samsung needs to apologize and reform immediately], *Hankyoreh,* October 13, 2016, http://www.hani.co.kr/arti/opinion/editorial/765552.html.

7 In 2019, the Supreme Court overturned the appeals court's decision, ordering Lee's retrial on corruption charges. For a historical overview of state-business relations, see Stephen Haggard and Chung-in Moon, "The State, Politics, and Economic Development in Postwar South Korea," in *State and Society in Contemporary Korea,* ed. Hagen Koo (Ithaca, NY: Cornell University Press, 1994), 51–94.

8 According to the brand consultant Interbrand, Samsung was ranked seventh among global brands in 2016. South Korea also sports a rapidly growing medical tourism industry appealing to travelers with innovative and reputable specialized treatments.

9 Representative studies include Yung Sik Kim, "Problems and Possibilities in the Study of the History of Korean Science," *Osiris* 13 (1995): 48–79; Geun Bae Kim, "An Anatomical Chart of South Korean Science and Technology in the 1960s," *EASTS* 5, no. 4 (2011): 529–542; Sungook Hong, "The Relationship Between Science and Technology in Korea from the 1960s to the Present Day: A Historical and Reflective Perspective," *EASTS* 6, no. 2 (2012): 259–265; Hee-Je Bak, "The Utilitarian View of Science and the Norms and Practices of Korean Scientists," in *Routledge Handbook of Science, Technology and Society,* ed. Daniel Lee Kleinman and Kelly Moore (London: Routledge, 2014): 406–418; Manyong Moon, "Understanding Compressed Growth of Science and Technology in South Korea: Focusing on Public Research Institutes," *Hanguk kwahak sahak hoeji* 37, no. 2 (2015): 431–453; and Sang-Hyun Kim, "Social Movements and Contested Sociotechnical Imaginaries in South Korea," in *Dreamscapes of Modernity: Sociotechnical Imaginaries and the Fabrication of Power,* ed. Sheila Jasanoff and Sang-Hyun Kim (Chicago: University of Chicago Press, 2015), 153–173.

10 Hiromi Mizuno, *Science for the Empire: Scientific Nationalism in Modern Japan* (Palo Alto, CA: Stanford University Press, 2008).

11 Sheila Jasanoff, "Future Imperfect: Science, Technology, and the Imaginations of Modernity," in Jasanoff and Kim, *Dreamscapes of Modernity,*

4. This definition is a refined version of an earlier definition by Jasanoff and Kim: "the collectively imagined forms of social life and social order reflected in the design and fulfillment of nation-specific scientific and/or technological projects." See Sheila Jasanoff and Sang-Hyun Kim, "Containing the Atom: Sociotechnical Imageries and Nuclear Regulation in the U.S. and South Korea," *Minerva* 47, no. 2 (2009): 119–146.

12 Jasanoff and Kim, "Containing the Atom," 123.

13 Jasanoff, "Future Imperfect," 4.

14 Fa-ti Fan, "Redrawing the Map: Science in Twentieth Century China," *ISIS* 98, no. 3 (2007): 524–538.

15 Hiromi Mizuno, "Introduction," *EASTS* 6, no. 1 (2012): 1–8.

16 Daniel Lee Kleinman, *Science and Technology in Science: From Biotechnology to the Internet* (Victoria, AU: Blackwell Publishing, 2005), 3.

17 Sang-yong Song, "Beyond Scientism: Coming of the Ethics of Science," *Hanguk kwahaksa hakhoeji* 35, no. 2 (2013): 389–398.

18 Michael Adas argues that wielding scientific and technological practices in Europe as measures of superiority and human value were deployed to justify ideologies of dominance in a time of European imperialist expansion in the nineteenth and twentieth centuries. See *Machines as the Measure of Men: Science, Technology, and Ideologies of Western Dominance* (Ithaca, NY: Cornell University Press, 1990). Dipesh Chakrabarty also critiques the rhetoric of modernity based on an evolutionary model as exemplified by Europe. See *Provincializing Europe: Postcolonial Thought and Historical Difference* (Princeton, NJ: Princeton University Press, 2000).

19 Fa-ti Fan, "Redrawing the Map," 527. Scholars such as Benjamin Elman and others address the complex dynamics of appropriation of STM in local contexts. See Benjamin Elman, *On Their Own Terms, Science in China, 1500–1900* (Cambridge, MA: Harvard University Press, 2005) and contributors to the international journal *East Asian Science, Technology and Society* (EASTS).

20 Francesca Bray lays out these perspectives in several publications. See *Technology and Society in Ming China (1368–1644),* American Historical Association–Society for the History of Technology, "Historical Perspectives on Technology, Society and Culture" pamphlet series no. 1, AHA, Washington, DC, 2000; "Technics and Civilization in Late Imperial China: An Essay in the Cultural History of Technology," *Osiris* 13 (1998): 11–33; and *Technology and Gender: Fabrics of Power in Late Imperial China* (Berkeley: University of California Press, 1997).

21 Yung Sik Kim calls this "the problem of China." In "Problems and Possibilities," he asks, "How should the historian deal with the theories and practices that were exactly the same as those of China except for the fact that they were discussed and carried out in Korea?" Kim republished several

essays, including this one, in *Questioning Science in East Asian Contexts: Essays on Science, Confucianism, and the Comparative History of Science* (Leiden: Brill, 2014).

22 Soyoung Suh, *Naming the Local: Medicine, Language, and Identity in Korea since the Fifteenth Century* (Cambridge, MA: Harvard University Asia Center, 2017).

23 Sang-woon Jeon, "Memories of the Early Days of the Korean History of Science Society, 1960–1980," *Hanguk kwahaksa hoeji* 33, no. 2 (2011): 349–358.

24 In his opus *Science and Civilization in China,* Needham showcases centuries of mathematics, inventions, and intellectual pursuits in Chinese early history, and asks, "Why did modern science, the mathematization of hypotheses about Nature, with all its implications for advanced technology, take its meteoric rise *only* in the West at the time of Galileo?" Nathan Sivin argues that Needham's implied why-not question, or why the scientific revolution did not develop in China (or India) despite their relative mathematical and scientific sophistication before the fifteenth century, misdirects historical studies of Chinese science. Sivin, "Why the Scientific Revolution Did Not Take Place in China—or Didn't It?" *Chinese Science* 5 (1982): 45–66. For a brief historiographical overview of Chinese science and technology, see Francesca Bray, "Science, Technology and Late Imperial History," *Chinese Historical Review* 24, no. 1 (2017): 93–104.

25 Yung Sik Kim, "The 'Why Not' Question of Chinese Science: Scientific Revolution and Traditional Chinese Science," *East Asian Science, Technology and Medicine* 22 (2004): 96–112.

26 Yung Sik Kim, "Problems and Possibilities."

27 For a historiographical overview of science in Chosŏn Korea, see Im Chongt'ae, "Chosŏn hugi kwahaksa yŏngu ŭi chŏnjaeng kwa kwaje" [Issues in the History of Science in the Late Chosŏn Period], *Yŏksa hakpo* 191 (2006): 449–463. In 1984, Seoul National University began a master and doctoral graduate program in history and the philosophy of science. Although not an independent department, the program does offer courses at the undergraduate level. As of 2017, the only STS undergraduate degree program in South Korea is in the Department of Science Studies at Jeonbuk (Chŏnbuk) National University in Jeonju (Chŏnju).

28 Yung Sik Kim, "Specialized Knowledge in Traditional East Asian Contexts," *EASTS* 4, no. 2 (2010): 179–183.

29 Fa-ti Fan, "Redrawing the Map," 527. Fan also cautions readers to be mindful that not all circulations are smooth or linear and may even be messy if not blocked. See Fa-ti Fan, "The Global Turn in the History of Science," *EASTS* 6, no. 2 (2012): 249–258.

30 Warwick Anderson, "Asia as Method in Science and Technology Studies," *EASTS* 6, no. 4 (2012): 445–451.

31 Jasanoff and Kim, "Containing the Atom." A recently published volume addresses connections between South Korea's socio-technical networks of the developmental era to those of the Japanese empire. See Hiromi Mizuno, Aaron S. Moore, and John DiMoia, eds., *Engineering Asia: Technology, Colonial Development and the Cold War Order* (London: Bloomsbury, 2018).

32 Max Fisher and Choe Sang-Hun, "How South Korea Flattened the Curve," *New York Times,* March 23, 2020, https://www.nytimes.com/2020/03/23/world/asia/coronavirus-south-korea-flatten-curve.html. See also Gregg Brazinsky, "South Korea Is Winning the Fight Against COVID-19. The US Is Failing," *The Nation Thailand,* April 10, 2020, https://www.nationthailand.com/opinion/30385757.

33 Sheila Jasanoff, "Imagined and Invented Worlds," in Jasanoff and Kim, *Dreamscapes of Modernity,* 337.

34 This volume does not address STM practices in North Korea. The field is relatively understudied; however, Korean STS scholars are currently turning their attention to North Korea. Recent publications include Kang Ho-Je, *Pukhan kwahak kisul hyŏngsŏngsa* [History of science and technology in North Korea] (Seoul: Sŏnin, 2007) and Dongwon Shin (Sin Tongwŏn), "Public Health and People's Health: Contrasting the Paths of Healthcare Systems in South and North Korea, 1945–60," translated by Seung-hee Jeon, in *Public Health and National Reconstruction in Post-War Asia: International Influences, Local Transformations,* ed. Liping Bu and Ka-che Yip (Oxford: Routledge, 2015), 90–111.

35 The phrase "locus of enunciation" comes from Warwick Anderson, "Asia as Method in Science and Technology Studies."

36 The *Journal of Korean Studies,* for example, published a special issue on science and literature in Korea to further understanding of reception of new sciences in Korea. See Dafna Zur and Christopher P. Hanscom, "Science and Literature in Korea: An Introduction," *Journal of Korean Studies* 23, no. 2 (2018): 213–222.

PART I

RECOLLECTING SOCIOTECHNICAL IMAGINARIES

CHAPTER 1

Aligning Patterns in the Material World

Sciences in Chosŏn Korea

Don Baker

Science is a modern conceptual construct. For most of the Chosŏn dynasty (1392–1910), Koreans did not use that term to distinguish thinking about natural phenomena from thinking about human society or moral issues. Rather than conceiving of science as a separate and distinct approach to conceptualizing or manipulating the world around them, they took a more holistic approach. Science in Chosŏn was more akin to natural philosophy than to modern science. It was more concerned with how human beings should harmonize with the natural world than with manipulating nature for human ends.

Studying nature in Chosŏn Korea entailed searching for patterns of appropriate interactions so that humans could act in accordance with those patterns and, by doing so, promote harmony in both nature and society. Sciences in Korea before the twentieth century were part and parcel of the neo-Confucian worldview, which viewed the realm of *ki* (C. *qi*), the physical world of matter and energy, through the lens provided by *li,* the intertwined patterns of appropriate interactions within the world of *ki.*[1] Modern scientists, on the other hand, separate the material realm (*ki*) and ethical values (*li*) in order to observe and analyze nature as it really is rather than approaching nature with moral assumptions predetermining what they should find. In other words, modern scientists are supposed to adopt an objective stance toward nature.

Neo-Confucians, on the other hand, because they viewed human beings as part of rather than apart from the material realm, defined "objectivity" differently. Theirs required cleaning their minds of any concern for personal benefit so that whatever they were observing could tell them how they should interact with it. That required viewing nature not as something existentially separate from the human realm but as defined by, and defining, human beings. What something really was, was determined by its appropriate interactions, including its interactions with human beings. In turn, natural objects were defined by the roles they played in nature as well as relative to human beings, and human beings were defined by the roles they played in nature as well as in society. A product of a search for patterns in nature that could provide clues to how, when, and why events occurred in the material world and thus could provide guidelines for appropriate ways to prepare for, prevent, or cope with those events, science in Chosŏn Korea produced behavioral, and therefore moral, guidelines.

Both the modern scientific and the neo-Confucian understandings of objectivity are problematic. Whether they are conducting experiments in their labs or observing natural objects in their natural state, modern scientists are looking for answers they have raised themselves rather than letting nature speak to them. This therefore leaves no way to escape an element of subjectivity in modern scientific practice. Similarly, even though neo-Confucians tried to remove considerations of personal benefit from their observations of natural objects and natural processes, they are nonetheless asking nature to tell them how they should behave, and that, too, creates a subjective framing of the questions they ask and therefore of the answers they find. Because neither approach can claim total objectivity, we should look for other criteria to understand science in Korea before the twentieth century.

One distinguishing feature of the earlier sciences was a focus on applicability rather than theory. Chosŏn sciences were not motivated primarily by abstract intellectual curiosity. Instead, they were predominantly a practical endeavor, driven by a need to uncover patterns of interaction within the natural world, such as those between heaven and earth, between the human body and the rest of the material world, and between energy within the earth and energy and fortune above it, which Chosŏn Koreans believed people needed to know in order to interact appropriately with and within their physical environment. Adopted from Chinese science to meet Korean ends in specific Korean environments, neo-Confucian science became Korean science.[2]

The patterns of interactions they observed attracted more attention than the physical objects actually interacting did. This pattern perspective was flexible and comprehensive enough to explain most natural phenomena Koreans were likely to encounter. However, it led to broad generalizations about how nature operated that did not often lend themselves to the generation of technology that would allow the manipulation of specific features of the natural world.

Moreover, those who thought systematically about the natural world were *yangban* who were supposed to be Confucian scholars of *li,* not *ki.* Tinkering with the *ki* realm was thought to be beneath their dignity. They usually left the invention of new technologies for the exploration or manipulation of nature to *chungin,* men who constituted an entirely separate, and inferior, social class of technical experts. Because of their status as the scholar-philosopher ruling class of the Chosŏn dynasty, *yangban* did not and could not serve as professional mathematicians, astronomers, geomancers, or physicians. Mathematicians, astronomers, geomancers, and physicians, on the other hand, as technical specialists did not wax theoretical on the implications of their work or on the philosophical assumptions that underlay it. Their task was not to discuss or debate philosophical approaches to conceptualizing the world they were measuring but instead to provide data their Confucian-scholar superiors could use. This split between philosophers and practitioners is as evident in mathematics as it is in the sciences that depended on mathematics in their practical applications.[3]

Mathematics

Koreans were accomplished mathematicians. From the start of the Chosŏn dynasty, they had the tools necessary for accurate adding, subtracting, multiplying, and dividing. Moreover, many of them knew how to wield those tools efficiently and productively. Those among them who had made a special effort to learn more advanced mathematical techniques could calculate the area of a square, a rectangle, a triangle, a circle, or even the volume of solids such as spheres. They were comfortable with both fractions and decimals and had available to them an approximation for the ratio of the circumference of a circle to its diameter (π) that was comparable in accuracy to the value used in the West. Moreover, they could work with negative numbers long before Europeans felt comfortable using them.[4]

Chosŏn mathematics was based on Chinese mathematics, which, during China's Song and Yuan dynasties that preceded the Chosŏn dynasty, surpassed European mathematics in many areas, particularly in developing algebraic solutions for higher degree equations and for the elimination of several unknowns in higher degree equations.[5] Koreans made no significant improvements or changes in the computational methods they learned from their Chinese neighbors. Nevertheless, Koreans used traditional mathematics during the Chosŏn dynasty to serve Korean ends and developed in a distinctively Korean milieu. For example, Koreans did not adopt the abacus, which had become widely used in China from the fifteenth century. Instead, they continued to rely on the computational rod notation found in the earliest Chinese mathematical classics.[6]

The Chosŏn government recognized that mathematics was an indispensable aid to governing the kingdom. Without the tools mathematics provided, officials knew they would be unable to determine how much arable land was on the peninsula and thus would be unable to predict how much tax could be collected from the grain that land produced. Nor, without the computations of competent mathematicians, would they be able to predict when solar and lunar eclipses would occur. That is why King Sejong (r. 1418–1450) established a Bureau for Studying Mathematics (Sŭpsan'guk) in 1423 and created thirty-nine official civil service positions for specialists in mathematics.[7] Most of those mathematicians were assigned to the Board of Taxation (Hojo), where the need for computational skills was most urgent. It was also the Board of Taxation that, along with the Board of Rites (Yejo), was given responsibility for administering the special civil service examinations used to select government mathematicians, a policy that continued for the rest of the dynasty.[8]

Mathematical techniques in Chosŏn were acquired by reading books from China, which focused on concrete solutions to specific problems rather than on general rules governing a class of problems.[9] For example, *Jiuzhang suanshu* (Nine Chapters on the Mathematical Art) had such chapter headings as "measuring fields," "cereals," and "making taxes fair." A computational formula introduced in Sino-Korean mathematical writings would be applied to a number of similar problems rather than abstracted into the form of a rule framed in symbolic universal terms. If that formula turned out to have wide applicability, then that applicability would be shown, not by restating it in the form of a theoretical principle, but by providing several more examples of that formula in operation. This is inductive reasoning, unlike the deductive reasoning European mathematicians have preferred since the time of Plato and Euclid.

The professional mathematicians of Korea who kept that approach alive were members of the *chungin* class of government clerks who had intermarried for centuries, creating a hereditary line of mathematicians stretching over four centuries. One examination of the roster of successful candidates for official mathematics posts from the end of the fifteenth century through the end of the nineteenth shows that of the 1,627 who passed the mathematics examination, all but 205 were the sons of other mathematicians.[10]

This, however, does not mean that the *yangban* ignored mathematics. In the second half of the Chosŏn dynasty, a few Confucian scholars expanded the range of their scholarly enquiries to include mathematics among the many subjects they wrote about. However, such *yangban* as Ch'oe Sŏkchŏng (1646–1715), Yi Ik (1681–1763), Hwang Yunsŏk (1729–1791), Chŏng Yagyong (1762–1836), and Ch'oe Han'gi (1803–1877) did not make mathematics their primary concern, nor did they make any significant contributions to the advancement of mathematics during the Chosŏn dynasty.[11]

The *yangban* scholars who dabbled in mathematics after 1600 had access to Western mathematical texts translated by Jesuit missionaries in China. Those texts were mostly introductions to Western approaches to arithmetic and geometry, showing an emphasis on geometry and its Euclidean logic. Although a few *yangban* read those books and adopted some of the techniques introduced in those Chinese-language publications, they did not change their basic approach to mathematics. They did not adopt the analytical and abstract approach Western mathematicians preferred, with its axioms and proofs. Instead, they retained the traditional Chinese and Korea preference for a concrete and contextualized approach.[12]

Astronomy

Throughout the Chosŏn dynasty, particularly after 1600, Korea neo-Confucian philosophers would occasionally stray from debates over metaphysics, ethics, ritual, or politics to discuss cosmological theories and the advantages and disadvantages of particular approaches to calendrical calculation, but normally they did not engage in calendrical calculation themselves. The technicians who had mastered the mathematical formulae necessary for calendrical calculation would, for their part, produce annual predictions of changes in the celestial patterns the sun, the moon, and the five visible planets created as they moved through the sky. Interaction between the theoretical concerns of the *yangban* and the practical concerns of the professional astronomers was minimal.

Yangban philosophers drew on the writings of Song dynasty neo-Confucians, in particular of Zhang Zai (1020–1077) and Zhu Xi (1130–1200), for their cosmological visions. No one single picture of the physical structure of the universe had been established to which all neo-Confucians had to subscribe. At times, Korean neo-Confucians seemed to prefer to think of heaven above as a sphere surrounding the earth like the white of an egg surrounds its yolk. More often, they envisioned the earth as a rounded bowl that is square at the base and has the sky as a hemispherical cover over it. In neither cosmological vision can we find any signs of the correlation of theoretical perspectives with experimental data that is the hallmark of modern science.

The *chungin* professional astronomers, on the other hand, limited themselves to providing maps of the skies and precise predictions of the movements of the sun, the moon, the planets, and the stars so that accurate ephemerides could be compiled. It was not their job to concoct or even evaluate theories that explained why the heavenly bodies moved the way they did. What shape the cosmos took was left for philosophers to decide. Sharing the mathematicians' concern for focusing on specific practical problems, Korea's astronomers concentrated more on predicting patterns of movement than on defining static properties of celestial objects. As long as they could use their mathematical formulae to calculate with a reasonable degree of accuracy where points of light in the sky would be from day to day, they did not worry about the specific physical characteristics of the arena through which those celestial bodies moved.[13]

Although Koreans might improve the astronomical instruments they borrowed from the Chinese and might even tinker with some peripheral Confucian cosmological ideas, few were bold enough to challenge the fundamental assumptions of the Sino-Korean neo-Confucian paradigm or the basic outlines of the picture of the cosmos it presented. They nevertheless turned Chinese astronomy into a tool to serve Korean ends, in this case the protection and enhancement of the authority and prestige of the Korean king.[14]

In the neo-Confucian worldview, the moral, the terrestrial, and the celestial worlds were intertwined. Events in heaven were believed to both reflect and affect events on earth. Astronomy entailed astrology. A solar eclipse darkening the daytime sky, for example, had more than purely astronomical significance. If predicted by the official calendar, that eclipse confirmed the monarch's claim to stand as the link between humanity and heaven and thus strengthened his right to rule. On the other hand, an eclipse that occurred without warning became an ominous portent proclaiming heaven's displeasure with the ruling house.[15] To maintain a secure grip on the throne, the Yi ruling family needed reliable notification of impending celestial events. A Royal Observatory (Sŏun'gwan) was in place from the first year of the Chosŏn dynasty. Although its name was changed in 1466 to the Kwansanggam, its prime function remained the same throughout the dynasty: strengthening the image of the king of Korea as the closest thing Korea had to a true "Son of Heaven."

The official Son of Heaven, residing in Beijing where he reigned as emperor, promulgated a calendar for all his subjects to follow, both those in China and those in neighboring tributary states. That calendar foretold only what Chinese would see when they turned their eyes skyward. It was not designed to forecast with equal precision what Koreans would see when they studied the stars in their sky. A perfunctory implementation of a Chinese calendar that did not take into account the changes in longitude and latitude from Beijing to Seoul could have brought more harm than good to the Yi family's claim to the throne. Total subservience to Beijing astronomers might strengthen the claim of the emperor of China to be the ruler of all under heaven, but in doing so it would underscore Korean dependency on China. Moreover, an unexpected eclipse or comet in Korea's skies could threaten the legitimacy of the Korean king. Unfortunately, when the government of China forwarded a copy of an official calendar to Korea, it did not include information on how that calendar was calculated. Koreans had to find out for themselves the formulae needed to adjust the calendar to Korean conditions. That is just what the fourth king of the dynasty, King Sejong (r. 1418–1450), ordered his astronomers to do.[16]

The result of that order was the publication of the *Ch'ilchŏngsan* (Calculations of the motions of the sun, the moon, and the five visible planets) in 1444. That manual was the first guide to calendrical calculation to be based on the latitude of Seoul rather than of the Chinese capital.[17] However, its existence had to be hidden from any Ming officials who might visit Seoul because the king of Korea was not supposed to usurp the prerogatives of the emperor of China by predicting celestial movements.

Nor could the Chinese be told of the advanced astronomical instruments that King Sejong had his technicians build for his Royal Observatory. The *chagyŏngnu* (automatic striking clepsydra), *ilsŏng chŏngsi ŭi* (sun-and-stars time-determining instrument), *kanŭi* (simplified instrument), *honŭi* (armillary sphere), *ilgu* (sun dials), and other devices the Koreans had constructed by improving on earlier Chinese instruments gave Korea what might well have been "the finest and most complete sets of astronomical instruments in the world" at that time.[18] Yet they, and the fact that Koreans were using their own observations to produce their own calendars, had to be kept secret.[19]

That secrecy had to be maintained for the remainder of the dynasty if Korea were to maintain the twin goals of good relations with China and accurate astronomical prediction at home. Two centuries after the secret publication of the *Calculations of the Motions of the Sun, the Moon, and the Five Visible Planets,* the Ming dynasty fell and the Qing dynasty that replaced it proclaimed a new calendar based on more accurate calendrical technology imported from Europe. Korea could not ask China's new rulers for the technology behind that Shixian (K. Sihŏn) calendar, for to do so would be to admit that Korea had been engaging in activities a tributary state should not engage in. Instead, Korea sent men to Beijing to discreetly learn the European methods of calendrical calculation.[20] That discretion, plus substantial quantities of silver placed in the right hands, enabled Korea to circumvent Qing restrictions and obtain enough information on Western ways of calculating the movements of the sun and the moon and other celestial objects to promulgate its version of a calendar based on Western techniques in 1653.[21]

Adopting the Western method of calendrical calculation did not mean adopting the Western conception of the structure of the cosmos. Most Koreans were not particularly interested in Western hypotheses about the actual physical placement of the various points of light they saw in the heavens.[22] The primary focus of Korean astronomers was on discerning patterns in the movements of those points of light. What exactly those points of light were was less important than how they interacted with each other. Moreover, those patterns were important not in themselves but instead for what they revealed about how human behavior on earth both influenced and was influenced by the behavior of celestial bodies.

Traditional Korean astronomy, based on the premise that everything in the universe was linked to everything else through the all-encompassing network of *li,* was inseparable from astrology. The king had to show his subjects that, thanks to his court astronomers, he was favored by Heaven with the ability to foretell heavenly movement, which gave him the power to tame potentially ominous portents. According to the *Kyŏngguk taejŏn* (National administrative code) promulgated in 1470, the staff of the Royal Observatory included not only twenty specialists in "celestial patterns" (astronomy) but also ten diviners, those qualified to interpret unexpected astronomical events as auspicious or ominous so that the king could respond to them by changing his behavior if necessary.[23] Astronomy was not studied for its own sake but instead for what the sky above had to say about the behavior of men below.[24]

King Yŏnsan (r. 1494–1506) was the only Chosŏn dynasty monarch to challenge the notion that unexpected celestial phenomena were indications that the person on the throne had not been acting properly and should rectify his behavior. In 1506, he ordered the officials in the Royal Observatory to suppress reports that a comet had unexpectedly appeared in the skies. When those officials reported that inauspicious event anyway, King Yŏnsan stripped the Royal Observatory of its responsibility for divination and ordered it to confine itself to the calendrical calculations only. Forty-two days later, he was overthrown, and diviners returned to the Royal Observatory, where they continued to work with astronomers for the rest of the Chosŏn dynasty.[25]

Geography

Serving alongside astronomers and diviners were the official geographers of the Chosŏn court.[26] Just as astronomers and diviners operated on the assumption that specific patterns in the skies could be directly correlated with specific patterns in human behavior and specific changes in human fortune, geographers operated on the assumption that specific topographical patterns on the earth could be directly correlated with specific patterns in human behavior and specific changes in human fortune.

Geography for most of the Chosŏn dynasty was inseparable from geomancy (K. *p'ungsu chiri*), the science of locating geographically auspicious locations. Maps drawn by geographers were often expected to show more than just where various mountains, rivers, and human settlements were located on the peninsula, or where Korea was located on the earth. Just as true of mathematics and astronomy, the patterns of relationships among and between the objects geography studied were more important that those objects themselves. Among the patterns geographer-geomancers were expected to uncover were those formed by the invisible, underground channels through which vital terrestrial energy (*chigi*) flowed. Their scientific investigations were supposed to locate the best places for human beings to tap into that energy flow.[27]

The theories and techniques of Korean geography and geomancy were borrowed from China. Their maps are normally accompanied by a text that points out what the geographic features depicted on the map meant to the human beings who lived on and around them.[28] How human beings experienced the land was more important than the actual physical layout. That is why most traditional Korean and Chinese maps indicate the presence of mountains by drawing a schematic version of what they look like viewed from the ground below rather than using the view from above common to modern Western cartography.

The oldest Korean map extant today, the *Honil kangni yŏktae kukto chido* (Integrated historical map of countries and capitals) of 1402, is a good example of the use of graphic representations of geography to make a point about human as well as spatial relationships. This is a map of the world as known to Koreans at the beginning of the Chosŏn dynasty. China dominates the map. Moreover, the Korean peninsula is given much more space on that map than its actual size would warrant. In fact, Korea appears much larger than Japan. Clearly, this map was intended to show that Korea recognized that China was the cultural center of the world, and that Korea was second in importance, and definitely more important than Japan, because it was culturally closer to China, and therefore more civilized, than Japan was.[29]

Further evidence that Korean geographers in general were as much concerned with cultural data as with spatial data are found in the popular, but unofficial, *ch'ŏnhado* (maps of all under Heaven).[30] These maps first began to be drawn in the seventeenth century, the same century Koreans first encountered world maps drawn by Europeans. However, they are very different from European maps of that era, and they are also unlike maps drawn in China and Japan at that time. Also known as wheel maps because they were round, they place a continental island dominated by China in the center, a few important islands such as Japan in the sea surrounding that island, and then another ring of land before the final, empty ring of water. Many of the numerous states identified on them were real political entities, but others were imaginary, taken from the ancient Chinese work the *Classic of Mountains and Seas* (Shanhai-jing) rather than from the real world.[31] Those maps, therefore, said less about the actual geography of the world than they said about the central cultural role China and, to a lesser degree, Korea played in that world.[32]

A third sign that Korean geography was concerned with more than simple physical topography lies in the way mountain ranges are usually drawn on traditional Korean maps. China was home to two primary schools of geomancy, one that focused on directions and used a compass and another that focused on forms and configurations and made its determinations of auspicious locations based on the shape and gestalt of mountains, hills, rivers and streams.[33] Korea was much more influenced by the "forms and configurations" school, and that influence shows in its maps.

Geomancers were interested in how the lay of the land channeled the flow of *chigi,* the energy of the earth. For example, over the course of the dynasty cartographers paid increasing attention to Mount Paektu, on the northern border.[34] It is depicted in maps, particularly in the latter half of the dynasty, as the source of the *chigi* that provided vitality for the Korean peninsula. In the eighteenth and nineteenth centuries, some private mapmakers began producing maps that displayed a greater concern for accuracy of detail and scale than previous maps had. That does not mean, however, that such geographers as Chŏng Sanggi (1678–1752) in the eighteenth century or Kim Chŏngho (1804?–1864?) in the nineteenth were concerned only with surface features. Their maps show graphically that rivers of energy flow from Mount Paektu and provide sustenance to all the mountain ranges of Korea. They treat the major rivers on the peninsula the same way. Both the flow of mountain ranges and the flow of river water are graphically related, either directly or indirectly, to the Mount Paektu headwaters of essential vitalizing energy.[35]

Mountains and rivers are important because they are the arteries and veins through which *chigi* flows. In the geomantic understanding of geography, the intimate relationship between human beings and the land on which they live is such that people can tap into the *chigi* flowing beneath the ground to enhance their well-being or that of their descendants.[36] However, not every location provides an optimal outlet for this energy. Geomancy was the science of locating which particular locations provided privileged access to that energy, both for graves for the dead and for buildings for the living. If a grave were located in a particularly auspicious spot, *chigi* would flow through the bones of those buried on that spot to their descendants, ensuring those descendants a successful and prosperous life. That was a core principle of geomancy, and one held so firmly and so widely that whenever a king or queen died, geomancers were consulted to ensure that the royal tomb was located in an auspicious site to protect the dynastic line.[37]

Royalty were not the only Koreans to try to assure an auspicious burial site for their loved ones. During the Chosŏn dynasty, many legal battles were fought over rival claims to particularly favorable burial sites.[38] More often than not, if a powerful *yangban* vied with a politically impotent *yangban* or with a commoner, the politically connected *yangban* won. His continued success and the later success of his children in passing the civil service examinations and winning appointments to government posts would then be offered as proof that burial sites did matter. If, on the other hand, his family fortunes declined despite his geomantic advantages, the family would be blamed for lacking the moral fiber to properly avail themselves of their privileged access to the energy of the earth, or for failing to move their ancestors' bones when the underground river of terrestrial energy shifted course.

Geomancy was used to determine auspicious locations for the living as well. For example, Seoul was chosen as the capital of Chosŏn because geomancers, after much discussion, decided it was the most auspicious location. Further, Kyŏngbokkung, the main palace, was placed at the most auspicious spot within the new capital city of Seoul to ensure that the dynasty would have maximum access to the vitalizing energy of the mountain that rose behind it.[39]

Yangban and commoners also relied on geomancy when selecting a site to build a home and when deciding how to orient that house once a site had been selected. Korea's first cultural geography, Yi Chunghwan's *T'aengniji* (Selecting a place to live), which first appeared in 1751, listed the geomantic conditions of a village, along with its economic potential, the presence of rivers and streams, and the character of the people who live there, as the most important factors to consider in choosing where to settle down and raise a family.[40] Moreover, residents of even geomantically well-endowed villages took special care to design their homes so that they faced the appropriate direction and had the appropriate arrangement of gardens and rooms to take maximum advantage of the terrestrial energy available in that particular location. In traditional Korean villages, the homes do not all face the same direction nor do they all have the exact same design. Geomancy required that homebuilders take into account the unique geomantic configuration of each house site.[41]

To the modern eye, the belief in invisible underground channels of energy that can influence the affairs of men, and that the best way to tap into the energy which courses through those channels is to build tombs, palaces, and homes in those places that provide the best access to that energy, does not appear scientific. However, during the Chosŏn dynasty, such an approach was considered just as scientific as those to astronomy adopted by the astronomers and the diviners who worked alongside the geographer-geomancers. All three approaches to nature were attempts to conceptualize the natural world in terms of natural rather than supernatural forces. In that sense, they can be called scientific.

Medicine

Traditional Korean medicine, although many of its concepts and its techniques lack the precise anatomical grounding essential to modern medicine, is equally scientific in the traditional sense of science as an attempt to understand the natural world in natural terms. For medicine, it was disease, injury, and death that needed to be understood, not astronomical anomalies or changes in family fortunes. Nevertheless, in their basic conceptual stance, physicians were no different from astronomers and geomancers or even mathematicians. They were more interested in patterns of interaction than they were in the actual material entities interacting.

In the traditional medicine of Chosŏn Korea, function preceded structure and physiology preceded anatomy. Korean physicians recognized the existence of bodily organs, of course. However, Chosŏn physicians felt little need for dissections to examine the actual physical structure of the other organs given that what those organs looked like was much less important than the functions they served. In the mid-eighteenth century, a Korean visitor to Japan was shocked to hear Japanese extolling the merits of the knowledge gained from cutting open the bodies of the recently dead. He told them, "Those who learn through dissecting are fools."[42] Some medical handbooks included a few illustrations of what the Koreans thought was the internal structure of the body. Those drawings, however, relied on descriptions in ancient books rather than observation.[43]

The anatomical terminology of traditional Korean medicine sounds familiar to those accustomed to Western anatomical terminology. The terms actually mean something quite different to someone trained in the Sino-Korean medical tradition, however. In fact, it would be more accurate to refer to them as "visceral systems of function" than as organs, given that the Sino-Korean names for those organs refer more to physiological functions than to anatomical structures.[44]

Korean physicians were not surgeons. Even in cases of malignant growths of the sort we today call cancer, instead of removing or refashioning a specific dysfunctional organ within a patient's body, they used pharmaceuticals, acupuncture needles, or moxibustion (the burning of small pieces of mugwort on or near an acupuncture point on the skin) to try to restore the entire body to proper physiological harmony.[45] Their goal was to preserve and promote a harmonious interaction of all the physiological and psychological activities within the body and between the body's internal operations and its external environment. That was their definition of health.

They assumed the most effective way to restore a patient's health was to correct a malfunction in the flow of *ki,* the animating energy within the body. Whether one organ or another was overly stimulated and generating too much *ki* or was sluggish and not generating enough *ki,* it could be restored to normal functioning through ingestion of medicinal material or stimulation with acupuncture or moxibustion. They saw no need for surgery. In fact, surgery, by altering the body, would cause more problems because surgical alteration would make it more difficult for the various parts of the body to interact normally, and therefore for the body as a whole to interact appropriately with its external environment.

Among the different approaches Sino-Korean medicine offered for treating physiological dysfunction, Koreans generally preferred "reinforcing" over "purging" for both the prevention and the treatment of disease. Korean physicians tended to see most illness as due more to internal weakness than to some invading pathological agent.[46] They therefore advised preventive medicine and the consumption of tonics such as ginseng more often than they prescribed emetics or other purgatives. This is the approach we see in the most famous medical work to come out of the Chosŏn dynasty, the *Tongŭi pogam* (Treasury of Eastern medicine).[47]

This emphasis on preserving and restoring physiological harmony was faithful to mainstream Chinese medical theory. In the terminology they used in diagnosing and treating a patient, Korean physicians could not be easily distinguished from their Chinese counterparts, who had a similar preventive and restorative perspective. Korean medicine differed from Chinese medicine, however, in the ingredients in its pharmaceutical prescriptions. Many of the plant, animal, or mineral components of Chinese prescriptions were not easily available in Korea. Korea traded with its Chinese, Japanese, and Ryukyuan neighbors to obtain some of those ingredients, but imported medicines tended to be expensive and in limited supply.[48] Those in need of them might not be able to afford them or, even if they could, might not be able to obtain them in time. Therefore, even before the Chosŏn dynasty, Koreans began Koreanizing its Sino-Korean medicine by searching for locally grown substitutes for expensive or rare foreign products.

Sometimes this search was led by *yangban* officials rather than by practicing physicians. Medicine, like mathematics, astronomy, and geography, was a field that all educated men were expected to have an amateur's acquaintance with. However, like practitioners of the other sciences, professional practitioners of medicine were not *yangban* but *chungin,* and, like mathematicians, astronomers, and geomancers, had to pass a civil service examination to qualify as professionals in their field. Belonging to a secondary status group, when working in government medical bureaus, they were limited to practicing medicine within the theoretical parameters established by their *yangban* superiors. No challenge to the dominant medical paradigm was permitted.

Nor were government-certified physicians allowed to set up a private practice. Their primary responsibility as government employees was the health and safety of the royal family and central government officials. Just as official *chungin* mathematicians, geomancers, and astronomers did not generally make their skills freely available to the public, neither did the physicians of the Chosŏn dynasty. In one aspect, however, Chosŏn physicians differed from their fellow *chungin* technical specialists: male physicians had female counterparts. Although no women were mathematicians, astronomers, or geographers, after 1406 women did practice medicine in Korea, though their status was lower than that of male physicians.[49] The reason lies in the nature of the medical profession and in the constraints of Confucian ethics.

Unlike other technical specialists, physicians had to touch other human beings in order to properly carry out the tasks assigned to them. That meant that a physician might have to come into physical contact with women who were ill, a violation of the Confucian insistence on a rigid separation of the sexes among upper-class adults. To ensure both the health of his consorts and respect for Confucian tradition, King T'aejong (r. 1400–1418) ordered that a few women be selected for training in administering acupuncture and moxibustion and in reading the pulse (an important diagnostic tool). Male pharmaceutical specialists could prepare medicines for upper-class women who were ill but could neither apply the acupuncture needle or moxa to their bodies nor place their male fingers on a female patient's veins to feel her pulse. Such tasks therefore were assigned to a small core of women medical practitioners at court.

Transformation in the Sciences

The neo-Confucian character of science in Chosŏn Korea remained largely unchanged over the course of the five centuries of that dynasty. That does not mean that Korean thinking about nature, how to understand it, and how to have the most productive relationship with it did not change. Koreans kept up with advances in scientific thinking and technology in China and made them their own, in some cases improving on them, as we can see in the Royal Observatory of King Sejong. In the second half of the dynasty, they also began acquiring via China some of the natural philosophy and technology of Europe. However, although they accepted star maps and calendrical calculation formulae with Western roots, they did not replace their traditional assumptions about how to approach mathematics, astronomy, geography, or medicine until the end of the nineteenth century.

Early conveyors to East Asia of Western Learning, as both European religion and European science and technology were called, were missionaries of the Society of Jesus: the Jesuits. One reason the Western scientific thinking those Roman Catholic missionaries brought to China in the seventeenth and eighteenth centuries did not undermine traditional scientific thought in either China or Korea was that it was not markedly superior overall in explanatory, manipulative, or predictive power to what Koreans already had available. Not as empirical, as experimental, or as mathematically oriented as later science, the Western Learning Koreans confronted in the seventeenth and the eighteenth centuries was conceptually dominated by the late medieval, early Renaissance version of Aristotelian natural philosophy.[50] The astronomy the Jesuit offered Koreans was pre-Copernican astronomy. Their mathematics was pre-Newtonian mathematics. Their medicine was pre-Harveyian medicine.[51]

The science the Jesuits introduced to East Asia did excel in certain areas. One of those was geography, particularly the geography of the globe. Not long after their arrival in China at the end of the sixteenth century, the first Jesuit missionaries in China, Frs. Matteo Ricci (1552–1610) and Michael Ruggieri (1543–1607), announced that "the earth is round and has inhabitants living all around it."[52] The Chinese, who believed that they lived in the center of the world, were at first surprised by this challenge to the popular East Asian belief that heaven is round but the earth was square.[53] Koreans, who received a copy of a Jesuit map of the earth as early as 1603, were just as surprised.[54] However, by the end of the seventeenth century, most knowledgeable Chinese and Korean scholars accepted the notion that the world was indeed round. That reevaluation of one of their core traditional beliefs did not lead to the transformation of the traditional approach to geography the Jesuit missionaries had hoped it would bring.

Korean geographers incorporated insights and information acquired from Jesuit maps into their own geographical tradition without radically altering that tradition in the process.[55] Long after their first exposure to Jesuit maps of the world, Koreans cartographers continued to depict mountain ranges in the traditional style highlighting their role as geomantic channels of terrestrial energy. Moreover, the *ch'ŏnhado* format for world maps remained popular, Western influence being detectable primarily in the names of some of the nations on the periphery.

We begin to see changes in the fundamental assumptions behind traditional Korean science in the second half of the eighteenth century, a century and a half after Koreans had begun reading books written in Chinese explaining European approaches to the natural world.[56] After becoming intrigued by the novelty of presentations of European natural philosophy and impressed by the fact that some of those European approaches to the natural world were being used in China, a few Korean Confucian scholars, notably Hong Taeyong (1731–1783), Pak Chiwŏn (1737–1805), and Chŏng Yagyong (1762–1836), at times shifted their gaze away from the intricate network of interactive processes that was the core of the neo-Confucian conception of the material world to examine individual objects as separate and distinct physical entities.[57] In the process of doing so, they moved closer to studying nature on its own terms rather than imposing assumptions of appropriate behavior onto unconscious natural objects and giving them moral import. They extracted the "is" from the "ought," a prerequisite for objective scientific observation of the natural world.[58] Also, by focusing on the various features of individual objects rather than defining them primarily in terms of how they interacted with things around them, they paved the way for the more analytical approach to individual natural objects characteristic of modern science.

In the nineteenth century, more up-to-date Western scientific thinking began to enter Korea via publications by Protestant missionaries from North America and the United Kingdom. Those works stimulated Ch'oe Han'gi (1803–1879) to go much further than his predecessors had and insist that empirical observation and measurement of individual objects was the surest route to accurate knowledge.[59] His was a lonely voice, however.

Not until the last quarter of the nineteenth century did modern science begin to seriously threaten the traditional worldview.

Korea was shaken out of its trust in traditional science and technology by changes in the international environment around it. The ease with which Western powers, equipped with modern technology, were able to defeat China in a series of battles in the nineteenth century undermined the Korean assumption that the Chinese approach to science and technology was superior to that of the West in most areas. The perception that Japan's use of modern science and technology helped it defeat China in the 1894–1895 Sino-Japanese War encouraged some Koreans to jettison traditional ways of thinking in order to survive in the newly competitive world order.[60] Unfortunately for Korea, that was too little, too late. Japan absorbed the independent kingdom of Chosŏn into its empire in 1910.

Although Korea began moving away from the traditional neo-Confucian approach to nature in the last couple of centuries of Chosŏn, at first those moves were tentative and slow. The shift from traditional science to modern science that began to accelerate under Korean leadership in the last decades of the nineteenth century gained momentum between 1910 to 1945, when Japan controlled education, occupational certification, and industrial development on the peninsula. By the time the Japanese were forced by their defeat in World War II to leave Korea, the traditional sciences had been discredited. The ways Koreans talk and write about nature, the ways they study it and try to manipulate and control it, and the ways scientists practice their craft, are dramatically different today than they were for most of the Chosŏn dynasty.

First, the hierarchical separation of theorists and technicians has faded. Korean universities today have schools of engineering and schools of medicine alongside schools of arts and letters. Those who try to understand how the material world operates now work alongside those who apply such scientific insights to creating new mechanical tools for exploring and manipulating natural objects. Sometimes they are even the same person. The old distinction between *yangban* and *chungin* has almost completely disappeared.[61]

So has the low status and gender discrimination of the way scientists were traditionally viewed. Mathematicians, astronomers, geographers, and physicians are now considered professionals, sharing high status with government officials and other elite members of Korean society. Moreover, not only can women work as physicians, they are given the same responsibilities as male physicians. In addition, although they are not yet represented in proportion to their percentage of the educated population, many women now work as mathematicians, astronomers, and geographers.

Not only has the traditional role of scientists in society disappeared, that of mathematicians and astronomers has as well. Chosŏn dynasty tools for numerical calculation and celestial observation have been replaced by more accurate methods from the West. Moreover, maps no longer reflect the mindset of premodern Koreans. Instead, maps now are drawn to reflect the actual physical measurements of the landforms and shapes reproduced on those maps.

Not all traditional sciences have disappeared. Geomancers are still frequently called upon to determine the best sites for buildings and graves. However, the selection of gravesites according to geomantic principles is not nearly as common as it used to be, partially because little hillside land remains for graves and therefore more and more Koreans are cremating their dead rather than burying them.[62] Traditional medicine has survived as well. In fact, it thrives alongside biomedicine. The large number of traditional medical clinics and even hospitals in Korean towns and cities today, along with the continuing Korean interest in *poyak,* restorative tonics based on traditional medical principles, is proof.[63]

Nevertheless, the new paradigm for science, emphasizing mathematical and empirical analysis and replicable experiments, dominates Korean approaches to the natural world in the twenty-first century. *Li* as the ordering force in the cosmos has been replaced by chemicals and biological processes, and *ki* as the material thus ordered has been replaced by molecules and atoms. The moral universe has become a material universe. Scientists in Korea today study nature for what they can learn about how nature behaves, and how that behavior can be used and modified to improve human material standards of living. They no longer study nature to better understand how human beings ought to interact harmoniously with each other and with nature.

How the scientific endeavor itself is conceived and practiced has been transformed. Science in Korea today is analytical because it slices the universe into separate and distinct physical entities to make it possible to analyze and manipulate those entities more precisely. It is mathematical because it assumes that material objects have specific numerical characteristics (in other words, can be weighed and measured), and that natural processes unfold in a way that can be defined mathematically. It is experimental because it assumes that when someone measures or manipulates a natural object, the result will be exactly the same as when anyone else measures or manipulates that object in the exact same way. None of these characteristics are seen in earlier studies of nature.

Chosŏn scientists, rather than focusing on the separate and distinct components of the universe, were interested in determining how those components interacted with each other within the all-encompassing cosmic network of such interactions. Although earlier Koreans weighed and measured objects and used mathematical formulae to predict the course of natural processes, they did not attempt to create mathematical formulae, such as the law of gravity, which applied across a wide range of phenomena, because they did not view the material realm as defined by invariable mathematical laws of universal applicability. Instead, mathematical descriptions of particular phenomena were seen as specific to those phenomena.

As they conceived the universe, each object was defined by the totality of its interactions, and no two objects had the exact same set of interactions, and therefore no two objects would react the same way to the same stimulus. That is the prime reason they were not interested in rigid experimental verification. Because they did not assume the regularity in the universe that underlies the modern understanding of nature, they had no expectation that a manipulation of a natural object by one person in one environment would have the same result as a manipulation of a similar object in a similar environment by someone else. Moreover, because they thought in terms of interactions rather than inert substances, they shared a recognition of the observer effect with modern physicists. They assumed that the very act of observing something—that is, measuring or manipulating it—changes it, and because every interaction is slightly different from every other interaction, the results of one such interaction will be at least slightly different from the results of every other. To them, that meant that whether an experiment could be replicated was irrelevant.

Similarly, because Chosŏn scientists did not share the concern of modern scientists for mathematical analysis and experimental verification, they did not feel any need to define their terms as precisely as terms are defined in modern science. Key terms in premodern science were too vague to permit the sort of measurements and manipulation that would make predictive accuracy possible. How can you measure *li* or *ki?* Yet predictive accuracy is a hallmark of modern science and technology. Modern scientists have dropped the use of such nebulous terms in favor of those that can be rigidly defined mathematically and physically so that they can use them to more precisely predict how natural objects will behave.

Yet just because the traditional sciences in Korea were different from the modern sciences we see in Korea and the rest of the world today, we should not dismiss them as nonscientific. Instead, we should recognize that they represent a different type of science, one that met the needs of the Korean people for more than a thousand years, before the world around them changed and changed Korea in the process. By conceiving of traditional Korean science as science, we can better appreciate the beliefs and values that not only allowed premodern Koreans to reduce uncertainties in their lives but also made it possible for Koreans to construct one of the most sophisticated cultures in the premodern world. When we acknowledge the many accomplishments of Koreans in past centuries, we should not leave out their many scientific and technological accomplishments, accomplishments made possible by the assumptions that directed their view of nature and their attempts to interact with it in productive ways.

Notes

1 *Li* determined what were and what were not appropriate interactions between human beings, between human beings and natural objects, and between natural objects alone. *Li,* not *ki,* defined how both human beings and natural objects should behave. For a comprehensive look at how neo-Confucians viewed the material realm, see Yung Sik Kim, *The Natural Philosophy of Chu Hsi (1130–1200)* (Philadelphia, PA: American Philosophical Society, 2000).

2 I agree with Yung Sik Kim's argument that science and technology in Korea should not be viewed as merely copies of Chinese science and technology. Instead, they should be viewed as developing within the larger Sinitic cultural sphere, to which Korea was an active contributor. See Yung Sik Kim, "The Problem of China in the Study of the History of Korean Science: Korean Science, Chinese Science, and East Asian Science," in *Questioning Science in East Asian Contexts: Essays on Science, Confucianism, and the Comparative History of Science* (Leiden: Brill, 2014), 239–252.

3 For a useful survey of research on premodern Korean science, see Yung Sik Kim, "Problems and Possibilities in the Study of the History of Korean Science," *Osiris* 13 (1998): 48–79.

4 For a survey of the calculating techniques available to mathematicians during the Chosŏn dynasty, see Chang Hyewŏn, *Sanhaksŏro ponŭn Chosŏn*

suhak [Mathematics in the Chosŏn dynasty as seen in books on calculation techniques] (Seoul: Kyŏngmunsa, 2007).

5 Joseph Needham, *Science and Civilization in China,* vol. 3 (London: Cambridge University Press, 1959), 81–102; and Yan Li and Du Shiran, *Chinese Mathematics: A Concise History,* trans. John N. Crossley and Anthony W.C. Lun (Oxford: Clarendon Press, 1987), 111.

6 Kim Yongun and Kim Yongguk, *Han'guk suhaksa* (Seoul: Yŏlhwadang, 1982), 281–301. On the invention of the abacus in China, see Li and Du, *Chinese Mathematics,* 184.

7 Chŏn Sangun, *Sejong sidae ŭi kwahak* [Science during the time of King Sejong] (Seoul: Sejong taewang kinyŏm saŏphoe, 1986), 58.

8 Kim Yongun and Kim Yongguk, *Han'guk suhaksa,* 168–169.

9 Karine Chemla, "Mathematics, Nature, and Cosmological Inquiry in Traditional China," in *Concepts of Nature: A Chinese-European Cross-Cultural Perspective,* ed. Hans Ulrich Vogel and Günter Dux (Leiden: Brill, 2010), 255.

10 The exceptions had fathers who were physicians (124), translators (75), or astronomers (6), also of *chungin* status. See Kim Yong Woon, "Pan-Paradigm and Korean Mathematics in the Chosŏn Dynasty," *Korea Journal* 26, no. 3 (March 1986): 31.

11 Kim and Kim, *Han'guk suhaksa,* 215–280; and Kim Yong Woon, "Pan-Paradigm and Korean Mathematics," 41–43.

12 Don Baker, "Impotent Numbers: Korean Confucian Reactions to Jesuit Mathematics," *Korean Journal of the History of Science* 34, no. 2 (2012): 227–256.

13 Shigeru Nakayama, *A History of Japanese Astronomy: Chinese Background and Western Impact* (Cambridge, MA: Harvard University Press, 1969), 68.

14 Joseph Needham, Lu Gwei-Djen, John Combridge, and John Major, *The Hall of Heavenly Records: Korean Astronomical Instruments and Clocks, 1380–1780* (Cambridge: Cambridge University Press, 1986), 1–15.

15 Park Seong-rae [Pak Sŏngnae], "Portents and Neo-Confucian Politics in Korea, 1392–1519," *Journal of Social Sciences and Humanities* 49 (June 1979): 53–117.

16 *Chŭngbo munhŏn pigo* [Official encyclopedia of Korea, revised] 1908, I: 5b.

17 Sang-woon Jeon [Chŏn Sangun], *Science and Technology in Korea: Traditional Instruments and Techniques* (Cambridge, MA: MIT Press, 1974), 79–82.

18 Needham, *Hall of Heavenly Records,* 16–94.

19 Park Seong-Rae, "Rise of Confucian Portentology in Korea," *Journal of Social Sciences and Humanities* 48 (December 1978): 44.

20 *Sillok* [Annals of the Chosŏn dynasty], Injo 23:12 pyŏngsin (1645).

21 *Sillok,* Injo 24:6 muin (1646); and Jeon, *Science and Technology in Korea,* 83.

22 Don Baker, "Jesuit Science Through Korean Eyes," *Journal of Korean Studies* 4 (1983): 207–239.
23 Kuksa p'yŏnch'an wiwŏnhoe, ed., *Han'guksa* [The history of Korea], vol. 10 (Seoul: T'amgudang, 1990), 122.
24 Park Seong-rae, *Portents and Politics in Korean History* (Seoul: Jimoondang, 1998).
25 Park Seong-rae, "Portents in Korean History," *Journal of Social Sciences and Humanities* 47 (June 1978): 51–52.
26 *Han'guksa,* 182.
27 Yi Hwa, *P'ungsuran muŏsin'ga* [What is geomancy?] (Seoul: Ihaksa, 2013).
28 Cordell D.K. Yee, "Reinterpreting Traditional Chinese Geographical Maps," in *The History of Cartography,* vol. 2, book 2, ed. J. B. Harley and David Woodward (Chicago: University of Chicago Press, 1994), 35–67.
29 Gari Ledyard, "Cartography in Korea," in Harley and Woodward, *History of Cartography,* 244–249; and Jeon, 281–284. For a reproduction of *Honil kangni yŏktae kukto chido,* see Han Young-woo, Ahn Hwi-joon, and Bae Woo Sung, *The Artistry of Early Korean Cartography* (Larkspur, CA: Tamal Vista Publications, 1999), 11.
30 Oh Sang-hak, "Circular World Maps of the Joseon Dynasty: Their Characteristics and Worldview," *Korea Journal* 48, no. 1 (Spring, 2008): 8–45.
31 For an English translation of *Shanhai-jing,* see Anne Birrell, trans., *The Classic of Mountains and Seas* (New York: Penguin Books, 1999).
32 Ledyard, "Cartography in Korea," 256–267; Yi Ch'an, "Chosŏn sidae ŭi chidoch'aek," in *Han'guk ŭi chŏnt'ong chiri sasang* (Seoul: Minŭmsa, 1992), 87–105.
33 John B. Henderson, "Chinese Cosmographical Thought: The High Intellectual Tradition," in Harley and Woodward, *History of Cartography,* 217; and Hong-Key Yoon, *The Culture of Fengshui in Korea: An Exploration of East Asian Geomancy* (Lanham, MD: Lexington Books, 2006), 29–30. Yoon argues that the difference between those two schools appears more in theory than in practice, and that geomancers used both approaches, though forms and configurations were given priority over what a compass said.
34 Yang Bo-Kyung, "Perceptions of Nature in the Chosŏn Period," *Korea Journal* 37, no. 4 (Winter 1997): 134–155.
35 Ledyard, "Cartography in Korea," 305–329.
36 Choi Chang-jo, "Study of How Koreans View and Utilize Nature," *Korea Journal* 32, no. 4 (Winter 1992): 34; and Oh Sang-Hak, "The Recognition of Geomancy by Intellectuals during the Joseon Period," *Review of Korean Studies* 13, no. 1 (March 2010): 121–147.
37 Yoon, *Culture of Fengshui,* 67–70, 119–126.

38 Ibid., 166.
39 Ch'oe Ch'ang-jo, *Han'guk ŭi p'ungsu sasang* [Korean fengshui philosophy] (Seoul: Minŭmsa, 1984), 221–240.
40 Ch'oe Ch'ang-jo, "Chosŏn hugi sirhakchadŭl ŭi p'ungsu sasang," *Han'guk munhwa* 11 (1990): 481. An English translation of *T'aengniji* is now available. See *A Place to Live: A New Translation of Yi Chung-Hwan's T'aengniji: The Korean Classic for Choosing Settlements,* translated by Inshil Cho Yoon (Honolulu: University of Hawai'i Press, 2018).
41 For a discussion of the impact of geomancy on the famous traditional village of Hahoe, see Sung-kyun Kim, *Winding River Village: Poetics of A Korean Landscape* (privately published in Seoul, 1988).
42 Shin Dongwon, "The Characteristics of Joseon Medicine: Discourses on the Body, Illustration and Dissection," *Review of Korean Studies* 13, no. 1 (Spring 2010): 150.
43 Shin Dongwon, "Korean Anatomical Charts in the Context of the Asian Medical Tradition," translated by Yuseok Kim, *Asian Medicine* 5 (2009): 186–207.
44 Nathan Sivin, *Traditional Medicine in Contemporary China* (Ann Arbor: Center for Chinese Studies, University of Michigan, 1987), 68, 124–133.
45 For example, see how Hŏ Chun, considered the best physician Chosŏn produced, advised dealing with swelling in the abdomen. Hŏ Chun, trans. Kim Namil, Cha Wung Seok, et al., "Part I: Internal Bodily Elements," *Donguibogam: Treasured Mirror of Eastern Medicine* (Seoul: Ministry of Health and Welfare, 2013), 442–448.
46 Kim Tujong, *Han'guk ŭihaksa* [The history of medicine in Korea] (Seoul: T'amgudang, 1981), 284.
47 Kim Ho, *Hŏ Chun ŭi Tongŭi pogam yŏn'gu* [A study of Hŏ Chun's Tongŭi pogam] (Seoul: Ilchisa, 2000).
48 An Tŏkkyun, *Sejong sidae ŭi pogŏn wisaeng* [Public health measures during the reign of King Sejong] (Seoul: Sejong taewang kinyŏm saŏphoe, 1985), 98–121.
49 Yŏ Insŏk, Yi Hyŏnsuk, Kim Sŏngsu, Sin Kyuhwan, Pak Yunhyŏng, and Pak Yunjae, *Han'guk ŭihaksa* [The history of medicine in Korea] (Seoul: Institute for Research into Medical Policy of the Korean Medical Association, 2012), 116–117.
50 Edward Grant, "Aristotelianism and the Longevity of the Medieval World View," *History of Science* 16 (1978): 93–106. Grant has Aristotelianism fading in 1650 or thereabouts. A study of Jesuit scientific publications by Steven James Harris shows its influence lasting for at least another century. See Steven Harris, *Jesuit Ideology and Jesuit Science: Scientific Activity in the Society of Jesus, 1540–1773* (PhD diss., University of Wisconsin, Madison, 1988).

51 Benjamin A. Elman, *On Their Own Terms: Science in China, 1550–1900* (Cambridge, MA: Harvard University Press, 2005).

52 From the preface to Matteo Ricci's map of the world, cited by Kenneth Ch'en, "Matteo Ricci's Contribution to, and Influence on, Geographical Knowledge in China," *Journal of the American Oriental Society* 59, no. 3 (1939): 327.

53 See, for example, the note in the *Siku quanshu zongmu tiyao* [Notes on the works listed in the *Siku quanshu* catalogue]: 106:67, on Sabastiano de Ursis, *Biao Du Shuo* [The gnomon]: "Talk that the earth was a small sphere shocked a great many people when first heard in China."

54 Yi Sugwang, *Chibong Yusŏl* [Miscellaneous essays by Chibong Yi Sugwang], 2:34b–35a.

55 Ledyard, "Cartography in Korea," 236. Ledyard notes that "Korea's shift to the styles and methods of Western cartography occurred only toward the end of the nineteenth century." Despite the leading role of some Jesuit missionaries in an official geographical survey of China in the early eighteenth century, Western geography had little impact there, either. See Cordell D. K. Yee, "Traditional Chinese Cartography and the Myth of Westernization," in *History of Cartography,* 170–202.

56 For a comprehensive survey of Chosŏn Confucian responses to Western Learning, see Yonsei Taehakkyo Kukhak yŏn'guwŏn, ed., *Han'guk sirhak sasang yŏn'gu* IV [Studies in Korea's practical learning] (Seoul: Hyean, 2005).

57 For an analysis of how much Chŏng Yagyong did, and did not, move away from the Confucian approach to nature, see Yung Sik Kim, "Science and the Confucian Tradition in the Work of Chŏng Yagyong," *Tasanhak* 5 (2004): 127–166.

58 Don Baker, "Seeds of Modernity: Jesuit natural philosophy in Confucian Korea," *Pacific Rim Report* 48 (August 2007): 1–16.

59 Unsok Paek, "The Empiricist's Progress," *Yugyo munhwa yŏn'gu* [Journal of Confucian philosophy and culture] 8 (2007): 231–260.

60 Park Seong-Rae, "The Introduction of Western Science in Korea, 1876–1910," *Korea Journal* 21, no. 5 (May 1981): 29–38.

61 A recent study of contemporary physicists argues that influence from the Confucian respect for those who work with the brains over those who work with their hands still lingers. Dong-won Kim, "The Conflict between the Image and Role of Physics in South Korea," *Historical Studies in the Physical and Biological Sciences* 33, no. 1 (2002):107–129.

62 Chang-won Park, "Funerary Transformations in Contemporary South Korea," *Mortality* 15, no. 1 (February 2010), 18–37.

63 Gil-soo Han, "The Rise of Western Medicine and Revival of Traditional Medicine in Korea: A Brief History," *Korean Studies* 21 (1997): 96–121.

CHAPTER 2

Medicine as a Virtuous Art in Chosŏn and Colonial Korea

Sonja M. Kim

On April 9, 1885, a new medical establishment opened its doors in Korea, treating twenty outpatients and indicating three cases for amputation on its first day.[1] Called Chejungwŏn (House of succoring the people), the hospital operated under the charge of Dr. Horace N. Allen, physician to the US Legation in Seoul and a missionary of the American Presbyterian Board; it signaled the transition from Sino-classical medical traditions to biomedicine based on anatomical understandings of the body and their related physiology. Chejungwŏn treated patients using techniques familiar in the West and began training Koreans in biomedicine, laying the foundation for the emergence of a new category of physicians. That full jurisdiction of Chejungwŏn moved to foreign missionaries in 1894 and that it survived throughout the Japanese colonial period (1910–1945) to become one among South Korea's foremost medical hospitals attest to narratives that situate Chejungwŏn as the origin and missionaries as agents of modern medical transformations in Korea.[2]

These accounts, however, contrast with scholarship that addresses Japanese physicians and public health activities in foreign settlement areas in treaty port cities such as Pusan after 1876, or Korean intellectuals who dabbled with medical ideas learned from Jesuit and Chinese texts.[3] Additionally, to Korean officials and perhaps even His Majesty King Kojong (r. 1863–1907), Chejungwŏn was simply one effort among many to reinvigorate the kingdom amid the turmoil of domestic tensions and external demands to engage in new forms of treaty and trade relations. To place foreign missionaries as paramount in Korea's medical modernization promotes what Shin Dong-won calls the Allen myth.[4] This myth not only foregrounds foreign missionaries' role in Korean medical history but also elides Korean agency in adapting foreign medical activities.[5] Problematically, it dichotomizes medicine as distinct and often polarized entities (for example, old-new, Western-Eastern, scientific-superstitious) and fails to provide nuanced understandings of Korea's medical transformations.

Rather than attribute origins or delineate modern from tradition, this chapter establishes continuities in thinking about medicine as a moral practice in relation to the political order between the Chosŏn (1392–1910) and Japanese colonial periods (1910–1945). It situates shifts at the turn of the twentieth century in a longer history to make visible the ways a moral calculus of Sino-classical concepts, ensconced in Chosŏn period institutions, became sutured on modernist practices in public sanitation and biomedicine. Principles that guided statecraft, exhorted certain kinds of human social relations, and explained workings of the natural world framed medicine as a virtuous art in Chosŏn. These principles continued in the late nineteenth century to shape the understanding and enforcement of bodily practices inspired by Western and Japanese models, resonating with what scholars of East Asian history of science and medicine note as complex negotiations between newly introduced and existing knowledge and organizations.[6]

The incorporation of values such as filial piety and Confucian benevolence in late nineteenth-century health activities redefined responsibilities of the state and individuals in the provision of and engagement in quality (however defined) care practices. The late Chosŏn state implemented regulations in the creation of a new public health administration intended to serve a self-strengthening agenda amid civilizing anxieties in a rapidly changing East Asian world. It also articulated the duties and obligations of subjects of a new Korea in novel citizenship projects.

Moreover, colonial-period discussion of medical ethics related to the provision of services, professional conduct, and provider-patient relationships in Korea frequently circulated the cultural idiom *insul.* Echoing the Confucian concept of benevolence as understood in Chosŏn, *insul* promoted idealized values in medicine attributed to Korea's past traditions.[7] *Insul* may have drawn from Chosŏn period values but was evoked by the Government-General of Korea (GGK) at the moment of Korea's colonization by Japan in 1910 to rationalize its health administration. This accounts in part for the facile incorporation of Japanese imperialist and assimilationist health interventions as well as the popularization of *insul* as a form of medical compassion.

Medicine as Moral Practice in Chosŏn

Foreign travelers during the late nineteenth century often described their impressions of Korea as dirty, filthy, foul, and diseased.[8] To these observers, medical care in Korea was altogether absent, superstitious at best, or barbaric at worst. Such castigations belie the gulf in understandings of the body and management of disease between these foreign visitors and their Korean hosts. When Protestant missionaries arrived in Korea in the late nineteenth century, elite medicine, centered at the court and denoted as *ŭi* (醫), derived its authority from a Sino-classical tradition, folk medicine drew from religious beliefs and everyday customs, and an expanding private medical sector with drug markets and itinerant healers populated the medical landscape.

Unlike popular beliefs that credited the cause of bodily ailments to gods or spirits, the Sino-classical Confucian medical tradition understood the dynamics of illness to be within the body and prescribed therapeutic interventions such as tonics, bodily manipulations, or lifestyle to affect those internal processes. According to the encyclopedic medical text *Tongŭi pogam* (Treasury of Eastern medicine, 1613), attributed to the famed court physician Hŏ Chun, the human body is the literal metaphor of the cosmos.[9] The heavens (symbolized by the head) and earth (feet) are connected by meridians through which the vitalities acquired at conception circulate.[10] Fundamental aspects of this cosmology include *qi* (K. *ki,* vital energy), the oppositional yet mutual pairing of *yin* and *yang* (K. *ŭm-yang*), and the dynamic sequence of the Five Phases (K. *ohaeng*). The correspondence system of the *yin-yang* pairing and the Five Phases accounts for all phenomena, including bodily processes, and affects *qi*'s manifestations in the body. Because life depends on, is animated by, and is depleted by the myriad ebbs and flows of *qi,* any disruptions to *qi* may result in disorder and even death. What catalyzed these disruptions then was the focus of debates in medical scholarship.

Sino-classical medicine extended beyond theoretical musings of the human body. It also provided the rationale and moral principles that guided medical study among the elite and services by the state. Notionally, medicine was a craft in Chosŏn, and state-employed physicians were categorized in the technically skilled *chungin* secondary status group. This discouraged the *yangban* elite from practicing medicine because they preferred achievement on literary-based civil service examinations that would confer higher-status official appointments.[11] However, medicine was privileged in a way that other *chungin* occupations such as law or accounting were not. The Chosŏn state in its first decades acknowledged close affinities between medical and Confucian scholarship.[12] For example, *The Veritable Records of the Chosŏn Dynasty* recorded in 1434 that medical skill requires proper investigation in the principles of *yin-yang* and the Five Phases because only those who understand these are able to diagnose disease and prescribe medications.[13] Accordingly, it did not separate proper and skilled medical practice from Confucian moral conduct and appointed Confucian scholars to its medical institutions.

This allowed for the growth of scholar-physicians (*yu'ŭi*) among the *yangban* elite in Chosŏn whose professed self-cultivation in scholarship and moral behavior valorized their medical knowledge and skills.[14] Many *yangban* intellectuals engaged in medical study as part of a broader Confucian curriculum. Those whose talents recommended their medical practice or scholarship for government service were called scholar-physicians, delineated from medical students and practitioners of the *chungin* secondary status.[15] For example, in 1723 King Kyŏngjong (r. 1720–1724) approved receiving consultations from *chungin* court physicians as well as scholar-physicians when he was not feeling well.[16] In 1811, Hong Uk'ho was differentiated from other court physicians because of his *yangban* background and was recommended to diagnose King Sunjo (r. 1800–1834) on the basis of his skillful medical techniques.[17] Chŏng Yagyŏng, one of the foremost *yangban* scholars of late Chosŏn and well versed in scientific and medical scholarship, was called to court when King Sunjo's son the crown prince was ill.[18]

In addition to treating patients, *yangban* literati penned medical treatises. Chŏng Tojŏn, the alleged architect of Chosŏn's political, legal, and financial institutions along Confucian ideological foundations and a close advisor to the kingdom's founder Yi Sŏnggye (King T'aejo, r. 1392–1398), compiled *Chinmaek togyŏl* (Illustrated instructions on pulse diagnosis, 1389).[19] Chŏng Yagyŏng wrote *Magwa hoet'ong* (Comprehensive treatise on measles, 1798). Two scholar-physicians worked with Hŏ Chun on the *Tongŭi pogam.* Even monarchs commissioned or compiled medical treatises themselves. *Ŭiyangnon* (Treatise on medicine, 1463) is attributed to King Sejo

(r. 1455–1468), and King Chŏngjo (r. 1776–1800), penned the two-volume medical text, *Sumin myojŏn* (Excellent advice to prolong life, c. 1799).

Medicine offered rulers and the elite a way to posture themselves as engaging in a broader category of Confucian moral practices. For this reason, I call medicine centered at the Chosŏn court a virtuous art. Medicine drew from natural cosmologies and the moral order of Confucian thought that framed medical study as part of Confucian erudition and guided physicians' moral conduct as self-cultivation. The early Chosŏn period's major medical text *Ŭibang yuch'wi* (Classification compilation of medical prescriptions, 1445) articulates clearly the links between medicine, Confucian education, and ethical behavior.[20] The sections "On the Education of the Great Physician" (*Non tae'ŭi sŭpŏp*) and "On the Absolute Sincerity of the Great Physician" (*Non tae'ŭi chŏngsŏng*) instruct physicians to be well versed in theories of cosmological correspondence and the Five (Confucian) Classics so as to know the Way of humaneness or Confucian benevolence. A proper physician devotes himself to medical studies, keeps a clear mind when diagnosing, does not partake in gaieties when his patient is in pain, refrains from boasting, and prevents greed or personal profit from affecting his practice. In short, *Ŭibang yuch'wi* idealized physicians as scholarly, learned, and upright moral beings, two central components of proper self-cultivation along Confucian paradigms.[21]

King Sejo's *Ŭiyangnon* echoes these moral exhortations by categorizing healers based on their moral capacities, from which their healing abilities arise.[22] It asserts that physicians with steady emotional and mental states instruct patients to remain calm even in critical moments, thus preventing serious illness in their patients. Improper greed, arrogance, inability to keep one's wits intact, or careless prescription of medication obstructed healers from providing relief and could even bring harm and death to their patients. Preserving one's morality was to be among the physician's highest priorities. Failure to possess proper moral behavior befitting a physician provided the grounds for censure and dismissal. For example, the *Veritable Records of the Chosŏn Dynasty* recorded that the court recommended the dismissal of scholar-physician Yi Kongyun because he was frequently absent, demonstrated lack of diligence and sincerity in his studies, and failed to respond appropriately to other medical officers.[23] For his display of arrogance when he entered the palace, lack of respect, and use of aggressive medicines that worsened conditions for which he was prescribing, Yi was accordingly punished.[24]

Reinforcing these exhortations was the principle of benevolent governance (*injŏng*) that encouraged rulers to aid the ill and destitute and was the basis on which the Chosŏn court established a central medical bureaucracy. Proper management and provision of medical care demonstrate good Confucian statecraft by ensuring the health of the king and his officials so that they may administer properly and by promoting the people's welfare, which was the monarch's moral duty to his subjects. Accordingly, the Chosŏn state centralized a bureaucracy that recruited, trained, and employed physicians and female medical healers (*ŭinyŏ*) in official institutions throughout the kingdom. At the apex of this medical bureaucracy stood the palace clinic, Naeŭiwŏn, which treated the royal family and capital elites. Chŏnŭigam (Directorate of medicine) oversaw *materia medica* (pharmacological products) for the kingdom and dispatched care providers when needed.[25] Limited care for commoners and relief measures during epidemics or other crises were handled by lower-level organs such as the Hyeminsŏ and Hwarinsŏ. Scattered outside the capital were *yakpang* (clinic or medicinal storeroom), which managed local medical and welfare services.[26]

The state also commissioned the publication of medical texts throughout the dynasty as part of Confucian governance. The *Tongŭi pogam,* according to its preface, was directed by the king in the tradition of sage leadership. Other state publications include collections of prescriptions to manage fever-related illnesses during epidemics, such as *Kani pyŏgonbang* (1525) and *Sinch'an pyŏgonbang* (1613), and instruction manuals on how best to survive famines.[27] To make more accessible medical knowledge among the populace in the aftermath of devastations in the wake of the Japanese invasions of the late sixteenth century, simplified medical texts written on childbearing, emergency first-aid, and measles by Hŏ Chun were published in the vernacular.[28]

In addition, Confucian virtues that guided human social relations such as filial piety (*hyo*), loyalty to the king (*ch'ung*), and wifely duty (*yŏl*) encouraged individual care for one's body and those of others. For example, elites in outlying regions formed medical associations such as the one in the Kangnŭng area in the latter half of the dynasty.[29] Association members in Kangnŭng procured medicinal ingredients and hired physicians, both lacking in the area. According to the association's contract, members served their parents in a moral display of filial piety by having medicines on hand should their parents become ill. The state also awarded exemplary subjects who devoted themselves to the care of their ailing parent, parent-in-law, or husband, and memorialized their achievements in the *Samgang haengsilto* (Illustrated guide to the three bonds).[30] King Chŏngjo's medical text *Sumin myojŏn,* written while he cared for his ailing grandfather King Yŏngjo (r. 1724–1776) for several years, reflects the linkage between medical erudition and care with the core Confucian value of filial piety.[31]

Filial piety extended beyond the care of parents to include the protection of and care for one's body and health. Such personal care includes elements related to a *yangsaeng* (Ch. *yangsheng*) tradition among the yangban elite.[32] *Yangsaeng,* literally "nourishing life," drew from Taoist techniques to elongate life by preserving or cultivating one's vital essence or promoting bodily *qi*'s processes and were aligned with self-cultivation practices. In short, *yangsaeng* practices aimed to optimize personal health and prevent illness through physical regimens, such as breathing and exercise, correspondence with seasons and environs, and moderation of dietary and lifestyle habits. Adherents attributed filial piety to *yangsaeng* regimens as a way to nurture the life and body bestowed by one's parents. The similarities between acts instructed by *yangsaeng* and modern regimes of hygiene later informed Koreans' evolving understanding of bodily practices at the turn to the twentieth century.

Benevolent Governance and a Hygienic Citizenry

Despite the close connection between Chosŏn elite medicine and principles in Confucian metaphysics, statecraft, and moral self-cultivation, ideals did not always translate into actuality. The hierarchical nature of the state medical bureaucracy centered its institutions in the capital for the royal family and the elite, and Sino-classical medical knowledge was the privy of a small educated and primarily male elite.[33] Quality medical care remained inaccessible for most of the population.[34]

After the 1876 opening of treaty-specified ports in Korea, discussions on health shifted among intellectuals and at court as Koreans engaged in new forms of diplomatic relations, participated in official observation missions abroad, and became familiar with health services and administration as practiced in Japan, the United States, and European countries. These encounters challenged medicine as it had been practiced. Ideals that framed medicine as a virtuous art, however, were reshaped rather than abandoned. They continued to inform expectations for health and medical services provided by the state, one's personal health or bodily practices, and the healer-patient relationship.

Interest in Western organization of medicine and public sanitation stemmed from a reevaluation of Chosŏn-period statecraft traditions in a quickly changing geopolitical environment. For example, Kim Okkyun, an advocate for a more radical schedule of reforms, prioritized health in his 1882 recommendations to the king for administrative attention.[35] In a project Ruth Rogaski calls hygienic modernity, *wisaeng* (Ch. *weisheng*, J. *eisei*), the neologism used to translate Western models of sanitation, signaled a departure from the term's earlier associations with individual health regimens enmeshed in cosmological understandings of *qi*.[36] Over time, *wisaeng* came to encompass bodily practices such as bathing and spitting, medical knowledge that drew from clinical, anatomical, and biological sciences, infectious disease control, vaccination, burial customs, and other activities to promote the overall health of the populace.

Associated with self-strengthening (*chagang*) and Meiji Japan–inspired civilization and enlightenment (K. *munmyŏng kaehwa*, J. *bunmei kaika*) agendas, *wisaeng*, however, did not sever Korea's health administration from earlier practices completely.[37] It overlapped with existing health-preservation behavior of *yangsaeng* (such as measures to fortify the body, reduce harm, and alter surrounding environs).[38] Occasionally the terms *yangsaeng* and *wisaeng* were conflated, thus making *wisaeng* more legible to Koreans.[39] Discussions that explained *wisaeng* as the prevention and treatment of disease, "the excellent method to allow the *paeksŏng* [the people] live long," linked individual health-seeking routines to prosperity of the country, kingly concern for his subjects' welfare, and later, nationalist pursuits of health.[40] As Pak Yŏnghyo's 1888 memorial to the throne explained, *wisaeng* performed collectively produces a populace that flourishes.

Supplication to the proper emotions a king should display toward his subjects as the foundation of benevolent governance continued to inform the state's provision of health and medical services. In an 1885 appeal, George Foulk of the US Legation reassured the Korean court that missionary Horace Allen's request to establish a new hospital was "highly commendable" with "purely unselfish motives towards benefitting His Majesty's subjects."[41] Exhorting the harmonizing of emotions to ensure proper relations between the monarch and his subjects in terms legible to the court, Allen added that the hospital would "be gratifying to His Majesty to see his people cared for properly in their distress, while it would undoubtedly still further endear the people to their monarch and elevate them in many ways."[42] Moreover, Allen called Chejungwŏn in English as "His Corean Majesty's Hospital."[43]

Demands for state provision of health services remained couched in similar terms. An 1896 editorial in the *Tongnip sinmun* exhorted the government to serve the people through health activities.[44] A few years later, another editorial repeated the message: "It greatly benefits the people if the government genuinely loves its people and seeks to strengthen the country by working to keep the people free of disease."[45] The state continued to name its health facilities with Chinese characters that conjure images of kingly benevolence such as *hye* (favor, kindness) and *che* or *je* (rescue, succor, relieve). Examples include Chejungwŏn, its initial name Kwanghyewŏn (House of extended favor), and the government hospital Kwangjewŏn (House of extended deliverance) during the Taehan Empire period (1897–1910).[46] That the Taehan government reestablished welfare institutions such as the Hyeminwŏn (House that favors the people) in 1901 further attests to state intentions that new medical institutions follow Chosŏn precedents.[47]

The state's operation of Chejungwŏn and other related health institutions suggests that they were to succeed, not replace, Chosŏn-period institutions. The court appointed Korean officials from the Foreign Office to Chejungwŏn. Some worked on site as gatekeepers and secretaries to keep records and report to the Korean commissioner.[48] Chejungwŏn regulations on patient fees and pay schedules stipulated that "all bills [are] payable only on recovery," which, according to Shin Dongwon, followed protocols of Hyeminsŏ during the Chosŏn period.[49] Additionally, Allen remarked in his diary that officers and medical students attached formerly to the Hyeminsŏ were to be transferred to Chejungwŏn.

Medical activities themselves and the power of the state to intervene directly in people's lives, however, changed. *Wisaeng* introduced new physiological conceptualizations of the body and vigilant routines of disinfection and cleanliness. Accordingly, it involved novel measures such as the regulation of slaughtered beef, repair of streets, and restrictions on shamanism and acupuncture. Rationalizing *wisaeng*'s utility for governing purposes, the court established a new sanitary police to enforce particular bodily practices and contagious disease regulations and implemented measures to standardize the medical profession by the twentieth century.[50] Self-strengthening desires that prioritized the well-being of the state, and by extension the nation in the face of encroaching imperialist interests, tempered the once privileged position of state provision of welfare services and medical care as moral practice. The onus of *wisaeng* health practices deemed proper became increasingly placed on individuals.[51]

Enmeshed with shifting conceptualizations of civilization, *wisaeng* normalized new bodily habits and linked them to citizenship projects. Citizenship here refers less to possession and exercise of rights associated with a liberal tradition and more to the cultivation and location of selves in line with newly envisioned state authority and a collective social order.[52] In the Korean context, the people's obligations and allegiance were often paramount. Citizenship transferred subjects' loyalty from the king to the nation, now represented by the monarch. Performed as duty, citizenship practice was likened to ethics or moral behavior. In other words, individual comportment and bodily routines promoted as *wisaeng* became moral obligations of members of the larger community and later national or imperial subjects. Health and welfare no longer remained merely a benefit or blessing received or given by the virtuous arts of the king or elite leaders.

Wisaeng discussions placed the responsibility of an infrastructure that supplied clean water, smallpox vaccination, improved street conditions, and medical training and institutions on the state; self-cultivation ideals became directed toward the systematic mobilizing or disciplining of people on a collective scale. These discussions instructed individuals in disease prevention as a form of moral practice. Applying the Confucian statecraft logic that linked individuals to their homes and outward to the country and ultimately the cosmos, *wisaeng* proponents asserted that a person's healthy habits by extension protect the family, neighbors, and the country.[53] They demanded cooperation from everyone, from the high to low, old to young, male to female, merchant to farmer. To think of or be concerned with only one's body is selfish because it fails to consider others.[54] Individuals should thus abide by public health regulations and keep themselves clean. Other stipulations made public urination and littering punishable.

According to an 1896 newspaper editorial, if the body has no disease, then it becomes more perfect. If the body is perfect, then one may perform one's work well. Once the body is complete in this fashion, one's thoughts become honest (*ch'ungsil*), in a progression likened to Confucian self-cultivation modes.[55] Moreover, *yangsaeng* notions that linked mind and heart to health persisted. The physical body develops according to the clothes one wears and food one eats. The mental state too has the potential to harm the body and threaten life.[56] Therefore, to remain healthy, one must avoid discouragement and stress. Rest both the body and mind. In a reversal of the *hyo* filiality-orientation of *yangsaeng* imperatives, one editorial retorted that if a parent loves his or her child, she or he will abide by the tenets of *wisaeng*.[57]

Wisaeng instructions were interwoven throughout school curriculum in the sciences, physical education, and morals or self-cultivation (*susin*) classes, a required part of elementary education through the colonial period. *Susin* primers emphasized the development of patriotism and moral character, forging an organic link between the health of the individual and citizens' duty to the new nation.[58] The 1907 textbook *Ch'odŭng yullihak kyogwasŏ* (Elementary ethics) opened with the chapter "Protecting Health," instructing that citizenship (to become a person of and fraternize with others in the country in which one resides) involves self-cultivation through the body, mind, and morals.[59] Whether one cares for ailing parents as an expression of filial piety, washes oneself, exercises, cleans one's home, or seeks a physician instead of a shaman, *wisaeng* reinforced modernist exhortations of personal care, cleanliness, and eventually, biomedicine.

Insul, an Imperialist Intervention

The incorporation of Korea in the Japanese empire as a protectorate in 1905 and formal colony in 1910 may have divested public health and medical administration from Korean hands, but moral imperatives embedded in medical practices persisted. The Japanese protectorate and colonial governments exhorted Korean subjects to internalize "Japanese-dictated norms of modern life," which Todd Henry calls civic morality (*kongdŏk*), in a grand program of civic assimilation.[60] Central to civic morality was individual adherence to hygienic tenets and strict compliance to health regulations, particularly disease control measures. Whereas Korean elites envisioned hygienic modernity for self-strengthening and nationalist purposes, Japanese administrators pursued it as a "powerful mechanism of colonial subjectification."[61] Ensuing health administration and campaigns, school curriculum, and print media discussions solidified links forged between hygienic practices and individual civic duties, on the one hand, and strength of the state, now Japanese empire, on the other. This produced an ironic conflation in the ways to achieve differing Japanese imperialist and Korean agendas.[62]

During the colonial period, another neologism informed by Chosŏn moral medical practices circulated. Formed by the Chinese characters *ren*

(K. *in*) for humaneness and *shu* (K. *sul*) for skill, technique, or method, *insul* came to acquire two meanings during the twentieth century. The first speaks more generally to the medical skills or healing techniques that treat injured or ill bodies. The second alludes to medical practice as a virtuous art, idealizing a code of ethics presented as originating in Confucian values of benevolence. During the Chosŏn period, however, writers often wrote the character *in* to stand alone.[63] Moreover, when written in combination as *insul,* the term did not necessarily indicate medicine but referred to broader benevolent statecraft practices, such as the pardoning of prisoners.[64]

Korea's colonial experience popularized imperatives embedded in *insul* in new ways. When Japan incorporated Korea in its empire, Terauchi Masatake, former resident-general now governor-general of Korea, deployed the language of Confucian benevolence to make the claim that the Japanese colonial administration prioritized medicine and public health as integral to Korea's recovery. He sympathized with the sick, deplored the inadequacy of Korea's medical system, and praised the achievements of the Taehan Hospital established by the Japanese Residency General of Korea in 1907.[65] He also outlined plans to extend the charity hospital system (*chahye ŭiwŏn*) to all provinces. A countrywide system of public hospitals with "physicians of repute" and "effective medicines," Terauchi insisted, would provide the people with the "*insul* of resuscitating lives."[66]

Insul used in this passage alluded to Sino-classical medical texts replete with characterization of medicine as a virtuous art such as "medicine saves the dying and heals the wounded" or "medicine is a humane technique."[67] Terauchi placed the Japanese colonial state as the benevolent actor, moved by compassion and possessing the material techniques and equipment to implement the virtuous art of medicine in Korea whose state lacked the means to do so. However, couching the health and medical administrative activities of the GGK as the conferral of "the [Japanese] emperor's favor" on his new Korean subjects, Terauchi's statements massage what Pak Yunjae argues was at stake. GGK health policies in Korea were intended foremost to assuage Korean resistance to the imposition of Japanese rule and to protect and promote the health of Japanese residents in colonial Korea.[68]

Institutionalizing the principle of *insul* in the colonial government's medical administration legitimized Japanese rule and reinforced idealized conceptualizations of medicine. The offering of free care at the public charity hospitals echoed the welfare functions of Chosŏn practices. Advertising imperial donations for medical activities supported efforts to assimilate and positioned Korean colonial subjects as recipients of the Japanese emperor's grace. In medical education, the ordinance for the Keijō Medical School (Kyŏngsŏng ŭihak chŏnmun hakkyo) stipulated the "*insul* of saving lives" as its guiding principle.[69]

The characterization of medicine as a compassionate practice, expressed by the rejoinder "Medicine is *insul*" (*Ŭi nŭn insul ira*), circulated among Korean physicians as well.[70] Informed by colonial period constructions of the *yu'ŭi,* the association of medicine and its corollary *insul* as a virtuous art solidified.[71] At the turn to the twentieth century, practitioners of Korean medicine may have begun to distance authority of medical knowledge from Confucian sources, but they maintained the vocation or practice of providing care to pursue the Way.[72] Biomedical physicians also crafted medicine as a venerable vocation. In the 1930s, Yu Sanggyu, for example, defined the essence of medicine or medical technique (*ŭisul*) as the practical application of knowledge based in the natural sciences to promote the health of humanity in the prevention and treatment of illness. To him, medicine was not only a scientific art based on medical knowledge but also a sacred art that placed the lives of humans as its objective or higher purpose. To execute properly, he urged, physicians need to harmonize their hands, hearts, and minds. Medicine could not be reduced to surgical skills alone.[73]

Insul also enabled an opening for women in a predominantly male profession that for the most part offered women only extremely limited opportunities for medical study or employment as physicians.[74] By the 1920s and 1930s, medicine shed its associations with female slaves and entertainers and became one of the few occupations deemed appropriate for educated women. Advocates for female physicians posited a match between the medical vocation and women's gendered nature, articulating the logic of *insul.* A 1927 journal article, for instance, argued that women with their "large capacity for benevolence and kindness" are well suited for the vocation of "*insul* succoring humankind."[75]

Regardless of stated intentions, however, the charitable services of the colonial government were not consistent or so benign. Over time, the GGK phased out the charitable functions of its public hospitals, changing the name from charity to provincial hospitals (*torip ŭiwŏn*) in the 1920s.[76] Sources indicate dissatisfaction with discriminatory practices that both Korean patients and medical personnel experienced at the public Japanese physician-dominated hospitals. Foreign Protestant missionaries postured themselves and their facilities as compassionate alternatives to what they portrayed as the impersonal, aggressive interventionist, and costly services of Japanese medicine, but they felt increasing financial and regulatory pressures that limited their operations.[77] As during the Chosŏn period, quality medical care, particularly biomedical therapeutics, was inaccessible to most of the population. Many relied on alternate therapeutics such as folk healing traditional medicine.[78]

In this context, *insul* emerged as a rhetorical strategy to highlight the plight of patients unable to access or receive care. More specifically, it signaled a growing criticism of the exorbitant fees hospitals and physicians charged patients and castigated what was presented as the abominable conduct of physicians. For example, writer Pak Insŏk lamented in 1928 that physicians lacked *insul* and abused their patients' trust by delaying treatment or refusing to treat after gauging their ability (or lack thereof) to pay.[79] Other vices, such as extracting money from wealthy patients by prescribing unnecessary medication or hospitalization, were said to become so common by the 1930s that they drew the attention of police authorities.[80] Such practices went beyond the question of profits to become an issue of human morality and virtue, reducing physicians to the level of pawn dealers or butchers.[81] The dignity of *insul,* it was stated, was what was at stake.[82]

Although colonial policies and economics of the medical system contributed to the inaccessibility of medical care for many, popular lore of *insul* tied to an imagined past remained. Nostalgic discussions characterized *insul* as a faded memory or museum piece, "disappearing like a rainbow."[83] Not all observers may have agreed, but by the end of the colonial period, *insul* was most commonly represented as being exhibited when a physician practiced medicine without considering remuneration. In 1941, O Kŭngsŏn, dean of the medical school affiliated with Severance Hospital, understood medicine as the most conscientious of occupations, guided by noble intentions and not by profits, and he urged his colleagues to manifest "the human affection and humane character" of *insul* in their vocation.[84]

Conclusion

The folding of the moral compass of benevolent governance and proper human social relations in new practices of public health and medicine may suggest a departure from the Chosŏn dynasty. As presented in this chapter, the underpinnings of the Korean state's health administration, individual health-seeking behavior, and medical knowledge at the turn to the twentieth century shifted from Sino-centric conceptualizations of civilization to those belonging to a modern era of nation- and empire-building. Nevertheless, the moral tenor of health administration and directives remained difficult to shake.

Moral imperatives embedded in the neologisms *wisaeng* and *insul* that circulated in emergent public health measures, schools, and print media help explain why and how Koreans found the universalist language of hygiene so alluring and powerful. These imperatives informed state provision of medical care and education, medical professionalization, and individual motivations. They also legitimated the application of science and biomedicine to the unprecedented (and sometimes violent) bodily interventions in health administration for utilitarian purposes and facilitated Koreans' incorporation in emergent citizenship and Japanese imperialist projects.

Moreover, circulated at the moment of Korea's colonization, *insul* drew on imaginations of past practices. Popularized as an idealized form of medical ethics, *insul* came to reference the desired or expected performance of what was perceived as Confucian values from care providers. It provided rhetorical strategies to critique the services rendered by physicians in colonial Korea. Although *insul* may no longer circulate in the same manner, the term continues to allude to an understanding of medicine as a mission of compassion, commonly used today to describe medical charitable activities.

Notes

Earlier versions of this chapter were presented at the University of Southern California (2013), Conference of the Asian Studies Association (2014), and the University of Hong Kong (2017).

1 Horace Allen's diary entry of April 10, 1885. Reprinted in Horace Allen, *Allen ŭi ilgi* [The diary of Horace Allen], ed. Kim Wŏnmo (Seoul: Dankook University Press, 1991), 462. Scholars use variable terms in reference to the last decades of the nineteenth century such as late Chosŏn, Hanmalgi (period of the late Han), and Taehan chegguki (period of the Taehan Empire). I have chosen to use Korea and Korean as general references unless to indicate the Chosŏn kingdom throughout the dynasty (1392–1910) and the period after the establishment of the Taehan Empire before Korea's colonization by Japan (1897–1910).

2 Examples of this narrative include Yi Kidŏk, *Han'guk kŭndae sŏyang ŭihak kyoyuksa* [The early history of modern medical education in Korea] (Seoul:

Ch'ŏngnyŏn ŭisa, 2008) and Yi Manyŏl, *Han'guk kidokkyo ŭiryosa* [A history of Christian medical work in Korea] (Seoul: Acanet, 2003). Severance Medical School and Hospital, which are affiliated with Yonsei University, claim Chejungwŏn as its predecessor.

3 Chejungwŏn was not the first hospital in the peninsula to offer Western-style therapeutics, as Japanese established hospitals in the port and foreign settlement areas of Pusan, Wŏnsan, Inch'ŏn, and the capital Hanyang (Seoul) opened by treaty after 1876. Miki Sakae, *Chōsen igakushi oyobi shippeishi* [History of Korean medicine and disease] (Osaka: Shibun chuppansha, 1962).

4 Horace Allen was summoned to the palace and saved the life of Min Yŏng-ik, the nephew of Queen Min, in the aftermath of the bloody 1884 Kapsin coup. His medical skills won the trust and support of the royal family. On the Allen myth, see Shin Dongwon, *Hoyŏlja Chosŏnŭl sŭpkyŏkhada: mom kwa ŭihak ŭi Han'guksa* [Cholera invades Chosŏn: History of the body and medicine in Korea] (Seoul: Yŏksa pip'yŏngsa, 2004), 224.

5 Today, Seoul National University Hospital traces its history to the Chosŏn government's pursuit of health reforms. This includes the granting of royal permission to Methodist missionary Robert S. Maclay to establish medical and educational work in Korea before the Kapsin coup, in addition to Chejungwŏn and later the government hospital Kwangjewŏn established in 1899. Maclay, however, returned to Japan. Kwangjewŏn was later incorporated by Japanese colonial administrations in their hospital system, eventually becoming the hospital affiliated with the Medical Department of Keijō Imperial University, the predecessor of Seoul National University.

6 Scholarship that address negotiation and accommodation of new knowledge include Bridie Andrews, *The Making of Modern Chinese Medicine, 1850–1960* (Honolulu: University of Hawai'i Press, 2015); Benjamin A. Elman, *On Their Own Terms: Science in China, 1550–1900* (Cambridge, MA: Harvard University Press, 2005); Ruth Rogaski, *Hygienic Modernity: Meanings of Health and Disease in Treaty-Port China* (Berkeley: University of California Press, 2004); Pak Yunjae, "Yangsaengesŏ wisaengŭro, kaehwap'a ŭi ŭihangnon kwa kŭndae kukka kŏnsŏl" [From *yangsaeng* to *wisaeng*, medical discourse of the Kaehwa group and construction of the modern state], *Sahoe wa yŏksa* 63 (2003): 30–50; and Soyoung Suh, *Naming the Local: Medicine, Language, and Identity in Korea since the Fifteenth Century* (Cambridge, MA: Harvard University Asia Center, 2017).

7 Shin Dongwon, "Han'guk ŭiryo yulli ŭi yŏksachŏk koch'al: ŭisa yulli kangnyŏng (1955–97) ŭi punsŏk ŭl chungsim ŭro" [History of medical ethics in Korea: Focused on analysis of medical codes and covenants], *Ŭisahak* 9, no. 2 (2000): 163–204.

8 Missionary physicians and travelers often wrote such accounts. O. R. Avison, "Disease in Korea," *Korean Repository* 4 (March 1897): 90–94. See also Todd Henry, "Sanitizing Empire: Articulations of Korean Otherness and the Construction of Early Colonial Seoul, 1905–1919," *Journal of Asian Studies* 64, no. 3 (August 2005): 639–675.

9 The version of the *Tongŭi pogam* consulted for this chapter is a facsimile reproduction accompanied by a vernacular translation by the Kug'yŏk *Tongŭi pogam* wiwŏnhoe (Committee to translate the *Tongŭi pogam*). Hŏ Chun, *Kug'yŏk Tongŭi pogam* [*Tongŭi pogam* with vernacular translation] (Seoul: P'ungnyŏnsa, 1966). According to Soyoung Suh, the masterful synthesis of Chinese classical medical texts and Korean local knowledge of the *Tongŭi pogam* receives "more scholarly and popular attention than any other premodern medical text in Korea" and remains relevant today in what contemporary Koreans call *hanŭihak* (Korean medicine), a system of medicine that claims a Sino-classical medical heritage distinguished from biomedicine. Soyoung Suh, *Naming the Local,* 2.

10 On the *Tongŭi pogam*'s description of the human body, see Kim Ho, *Hŏ Chun ŭi* Tongŭi pogam *yŏngu* [Study of Hŏ Chun's *Tongŭi pogam*] (Seoul: Ilchisa, 2000); and Shin Dongwon, Kim Namil, and Yŏ Insŏk, *Hangwŏnŭiro ingnŭn* Tongŭi pogam [Reading the *Tongŭi pogam* in one volume] (Seoul: Tŭllyŏk, 1999), 18–44.

11 Chosŏn Korea was a status-based society, awarding higher bureaucratic appointments and privilege to the *yangban.* Qualifications for the highest levels of civil service examinations were restricted and hereditary. *Sŏŏl* ("illegitimate" sons of nonprimary wives) and sons of remarried widows, for example, were affected by these restrictions. The secondary status *chungin* group held lower bureaucratic eligibility and served in fields such as translation, law, accounting, painting, and medicine on *chapkwa* or technical examination. See Kyung Moon Hwang, *Beyond Birth: Social Status in the Emergence of Modern Korea* (Cambridge, MA: Harvard University Asia Center, 2004).

12 In the latter half of the dynasty, regional promotions were offered more frequently to physicians than other *chungin* occupations. Kim Yangsu, "Chosŏn hugi chŏnmunjik chungin ŭi Cholla chibanggwan chinch'ul," *Yŏksa wa sirhak* 17–18 (2000): 369–412.

13 *Sejong sillok,* 16/7/25 (1434). I use a digitized online source for the *Veritable Records of the Chosŏn Dynasty* (http://sillok.history.go.kr) by the National Institute of Korean History. For each record, I indicate the reign year of the monarch, month, and day, in that order. To facilitate reading, I also indicate the year according to conventional standards in parentheses.

14 We see a similar growth of scholar-physicians (C. *ruyi*) in imperial China. Volker Scheid, *Currents of Tradition in Chinese Medicine: 1626–2006* (Seattle, WA: Eastland Press, 2007), 33–58.

15 Scholar-physicians often worked in the Ŭiyak tongch'amch'ŏng (Office of medical participants) or Ŭisŏsŭp tokkwan (Office of medical texts). Kim Ho, *Hŏ Chun ŭi* Tongŭi pogam, 153; and Yi Kyugŭn, "Chosŏn sidae ŭiryo kigu wa ŭigwan" [Chosŏn period medical bureaucracy and medical officers], *Tongbang hakji* 104 (1999): 95–161.

16 *Kyŏngjong sillok,* 3/4/1 (1723).

17 *Sunjo sillok,* 11/9/5 (1811).

18 Shin Dongwon, *Hoyŏlja,* 184.

19 *T'aejo sillok,* 7/8/26 (1398).

20 King Sejong (r. 1418–1450) commissioned the *Ŭibang yuch'wi,* which discussed, annotated, revised, and categorized medical prescriptions from various Chinese medical texts. The first edition took three years to complete, referencing texts from the Han, Tang, Song, and Yuan Dynasties and amounted to 365 volumes. *Sejong sillok,* 27/10/27 (1445). *Ŭibang yuch'wi* underwent revisions and was republished in 1477. I consulted Kim Tujong's preface to the facsimile reproduction. *Ŭibang yuch'wi* 1 (Seoul: Kŭmyŏng ch'ulp'ansa, 1977). English translation for *Ŭibang yuch'wi* comes from Soyoung Suh, *Naming the Local,* 29.

21 Such moral exhortations do not surprise: Shin Dongwon notes that *Ŭibang yuch'wi* was adapted from the Chinese medical classic *Bei ji qian jin yao fang* by Tang physician Sun Simiao. Shin Dongwon, "Chosŏn sidae ŭi ŭihangnon" [Medical theory during the Chosŏn period], *Ŭisahak* 13, no. 1 (2004): 134–145. On the *Bei ji qian jin yao fang,* see Paul Unschuld, *Medical Ethics in Imperial China* (Berkeley: University of California Press, 1979), 26–27.

22 *Sejo sillok,* 9/12/27 (1463).

23 *Kyŏngjong sillok,* 4/4/28 (1724).

24 *Yŏngjo sillok,* 1/8/20 (1724).

25 Patients were not treated within a medical facility like hospitals of today but instead in their own rooms or homes by medical officers dispatched by the *Naeŭiwŏn.* The *Naeŭiwŏn* itself housed administrative offices, libraries, and pharmaceutical storerooms necessary to its work. Shin Dongwon, *Hoyŏlja,* 177.

26 Ibid., 191–198.

27 Kim Ho, "1612 nyŏn onyŏk palsaeng kwa Ho Chun ŭi *Sinch'an pyŏgonbang*" [Joseon's response to the 1612 plague—examination of Heo Jun's *Shinchan Byeong-onbang*], *Chosŏn sidae sahakpo* 74 (2015): 307–333; and Yi Kyŏngnok, "Chosŏn Chungjong 19–20 nyŏn ŭi chŏnyŏmbyŏng

ch'anggwŏl kwa kŭ tae'ŭng" [The epidemic outbreak under Chungjong's reign and its responses], *Chungang saron* 39 (2014): 5–40.

28 Hŏ Chun's vernacular medical texts include *Ŏnhae t'aesan chibyo, Ŏnhae kugŭpbang,* and *Ŏnhae tuch'ang chibyo.*

29 "Yakkye ipŭi," as cited in Shin Dongwon, "Chosŏn sidae chibang ŭiryo ŭi sŏngjang: kwan chudo esŏ sajok chudoro, sajok chudo esŏ sijang chudoro—Kangnŭng yakkye (1603–1842) ŭi chojik kwa haesorŭl chungsimŭro," *Han'guksa yŏngu* 135 (2006): 6.

30 *Yŏl* is often translated as female chastity, but this narrows the range of activities (such as caring for sick in-laws) for which women were awarded as exemplary behavior. I thus translate it as wifely duty.

31 Kim Ho, "Chŏngjo tae ŭiryo chŏngch'aek," *Han'guk hakpo* 22, no. 1 (1996): 1231–1257.

32 Kim Ho, *Hŏ Chun ŭi* Tongŭi pogam *yŏngu,* 150–173.

33 Kim Ho, "18 segi huban kŏgyŏng sajok wisaeng kwa ŭiryo," *Sŏŭl hak yŏngu* 11 (1998): 113–144.

34 Commoners received limited care from state medical institutions, often during critical moments of natural disasters, famines, or epidemics. Growth of pharmaceutical markets, medical text publications, and healers in the late Chosŏn period suggests an increasing familiarity with the *materia medica* and theories of scholarly medicine, even if many commoners found medicinal ingredients to fulfill prescriptions difficult to procure because of cost or availability, or partook in other forms of healing.

35 "The most vital policy that people seek in each country is first, *wisaeng,* second agriculture and commerce, and third transportation and roads." Kim Okkyun, "Ch'ido yangnon," reprinted in *Han'guk ŭi kŭndae sasang: Han'guk sasang chŏnjip* 6 (Seoul: Samsŏng ch'ulp'ansa, 1984), 87–95.

36 Rogaski, *Hygienic Modernity,* 16.

37 Self-strengthening refers to a broad category of reformist impulses that sought to maintain Sino-classical values and worldviews while selectively adapting material goods and techniques from the West. They drew their rationale from classical texts and had various appellations including *tongdo sŏgi* (Eastern Ways, Western Machines), *ch'e-yong* (Essence and Practical Application), and *kubon sinch'am* (Old Foundation, New Addition). This is not to argue that modernizing desires for self-strengthening or civilization and enlightenment were divorced from classical or Confucian worldviews but to make aware that these worldviews no longer dominated the principles that guided health reforms and administration.

38 Pak Yunjae, "Yangsaengesŏ wisaengŭro."

39 Pak Yŏnghyo's 1888 memorial explained *wisaeng* as "the *yangsaeng* that is spoken of by people from a civilized country." A text of Pak Yŏnghyo's

memorial *Pak Yŏnghyo kŏnbaeksŏ* may be found in Chŏn Pongdok, *Han'guk kŭndaebŏp sasangsa* (1981): 159–161.

40 "Manguk wisaenghoe," *Hansŏng sunbo,* May 5, 1884.

41 Letter from George C. Foulk to Cho Peung Ho, president of the Foreign Office, of the Foreign Office, January 27, 1885, reprinted in Allen, *Allen ŭi ilgi,* 429.

42 Letter from Horace N. Allen to Cho Peung Ho, president of the Foreign Office, January 27, 1885, reprinted in Allen, *Allen ŭi ilgi,* 431.

43 Chejungwŏn was initially named Kwanghyewŏn but was renamed as Chejungwŏn within a few days of its opening. *Kojong sillok,* 22/3/12 (1885).

44 *Tongnip sinmun,* May 2, 1896.

45 Ibid., July 19, 1899.

46 Allen, *Allen ŭi ilgi,* 463.

47 Activities of the *Hyeminwŏn* for the year 1902 were actively reported in *Hwangsŏng sinmun.* The institution was dissolved, however, the following year.

48 Allen, *Allen ŭi ilgi,* 460.

49 Shin Dongwon, *Hoyŏlja,* 226. This fee schedule is different than Allen's expectations to be compensated for services when rendered, regardless of outcome.

50 On public health initiatives implemented by the state, see Pak Yunjae, *Han'guk kŭndae ŭihak ŭi kiwon* [The origin of the Korean modern medical system] (Seoul: Hyean, 2005); and Shin Dongwon, *Han'guk kŭndae pogŏn ŭiryosa* [History of health and medicine in modern Korea] (Seoul: Hanul Academy, 1997).

51 Burden of acceptance or compliance expected of or placed on individuals include smallpox vaccinations, biannual household clean-ups, payments to the Seoul Sanitation Association, and compulsory isolation and quarantines (Henry, "Sanitizing Empire"). Shin Dongwon and Hwang Sangik argue that this is not the rise of the modern individual according to a liberal tradition but a subject of a state that is benevolent at the same time its measures are coercive. Shin and Hwang, "Chosŏn malgi (1876–1910) kŭndae pogŏn ŭiryo ch'eje ŭi hyŏngsŏng kwajŏng kwa kŭ ŭimi" [Formation of the modern health and medical system in late Chosŏn], *Ŭisahak* 5, no. 2 (1996): 155–165.

52 Kyung Moon Hwang, *Rationalizing Korea: The Rise of the Modern State, 1894–1945* (Berkeley: University of California Press, 2015).

53 *Tongnip sinmun,* May 19, 1896. For an expanded discussion of health directives in civic education, see Sonja M. Kim, "The Search for Health: Translating *Wisaeng* and Medicine during the Taehan Empire," in *Reform and Modernity in the Taehan Empire,* ed. Kim Dong-no, John B. Duncan, and Kim Do-hyung (Seoul: Jimoondang, 2006), 299–336; and *Imperatives*

of Care: Women and Medicine in Colonial Korea (Honolulu: University of Hawai'i Press, 2019).

54 *Tongnip sinmun,* May 9, 1896.

55 Ibid., May 2, 1896.

56 Ibid., September 29, 1899.

57 Ibid., June 27, 1896.

58 On education and citizenship in late nineteenth-century Korea, see Hwang, *Rationalizing Korea*; and Leighanne Kimberly Yuh, "Moral Education, Modernization Imperatives, and the *People's Elementary Reader* (1895): Accommodation in the Early History of Modern Education in Korea," *Acta Koreana* 18, no. 2 (2015): 327–355.

59 An Chonghwa, trans., *Ch'odŭng yullihak kyogwasŏ* [Elementary ethics textbook] (Hansŏng: Kwanghak sŏp'o, 1907), 1–2.

60 Todd Henry, *Assimilating Seoul: Japanese Rule and the Politics of Public Space in Colonial Korea, 1910–45* (Berkeley: University of California Press, 2014).

61 Ibid., 130.

62 This conflation explains the betrayal Korean self-strengtheners faced after Japan proclaimed Korea as a protectorate. Andre Schmid, *Korea Between Empires, 1895–1919* (New York: Columbia University Press, 2002).

63 I thank Dr. Hwang Chaemun for alerting me that the Chinese characters *in* and *sul* were not commonly used in conjunction before the modern era.

64 An example of this is recorded in *T'aejong sillok,* 14/6/9 (1414).

65 The protectorate government incorporated the Chosŏn government hospital Kwangjeŏn into its Taehan Hospital, the predecessor of the GGK Hospital, and in the postcolonial period, Seoul National University Hospital.

66 Terauchi Masatake's message to the Korean people was recorded in *Sunjong sillok,* 3/8/29 (1910).

67 Shin Dongwon explains that although Chosŏn period medical text texts other than *Ŭibang yuch'wi* did not articulate medical ethics, Chinese texts that did were widely read in Chosŏn. He notes that late Chosŏn period physician Hwang Toyŏn expressed medicine as an art of *ren* in his late nineteenth-century medical texts. Shin Dongwon, "Han'guk chŏnt'ong ŭihak ŭi ŭihak yulli wa saengmyŏng yulli: Chosŏn sidae ŭi ŭisŏrŭl chungsimŭro" [Bioethics and medical ethics in Korean medical tradition], *Chonggyo munhwa pip'yŏng* 5 (2004): 65–90.

68 Pak Yunjae, *Han'guk kŭndae.*

69 As cited in Shin Dongwon, "Han'guk ŭiryo yulli ŭi yŏksachŏk koch'al," 177.

70 An example of this phrase appears in the caricature piece "Sisang manhwa" [Caricature of the times], *Pyŏlgŏngon* 33 (1930): 60–64.

71 Pak Yunjae, "Singmiji sigi ŭisa kyech'ŭng ŭi sŏngjang kwa chŏngch'esŏng hyŏngsŏng" [Identity formation and development of a physician class during

the colonial period], in *Tong Asia yŏksa sok ŭi ŭisadŭl* [Medical doctors in the history of East Asia], 150.

72 Soyoung Suh, "Korean Medicine Between the Local and the Universal: 1600–1945" (PhD diss., University of California, Los Angeles, 2007). Suh also describes the creative ways such practitioners, licensed yet restricted by the GGK as "students of medicine" (*ŭisaeng*) in contrast to licensed biomedical physicians (*ŭisa*), carved a space for their medical knowledge and practice by "provincializing biomedicine." Suh, *Naming the Local.*

73 Yu Sanggyu, "Seburansŭ pyŏngwŏn munje rŭl chungsimŭro" [On the Severance Hospital problem], *Sin Tonga* 5, no. 4 (1935), reprinted in Yu Sanggyu and Yu Yongsŏp, *Aeguk chisa T'aehŏ Yu Sanggyu: chŏngi wa yugo* [Patriot T'aehŏ Yu Sanggyuk: His life and works] (Seoul: Hongsadan ch'ulp'anbu, 2007), 253–263.

74 The government medical schools and Severance Medical College run by foreign missionaries did not accept female students. Because no institutions with full accreditation accepted women to study medicine in Korea until 1938, women studied medicine abroad or qualified for a medical license by passing a licensing examination after studying independently or at a training institute which started in the late 1920s. Although official numbers of female physicians were not recorded, discussions among physicians published in print media estimate between thirty and fifty in the 1930s. Sonja Kim, *Imperatives of Care.*

75 "Yŏja chigŏp annae, ton ŏpsŏsŏ woeguk yuhak motgago ch'wijikhal kosŭn myŏtchina toenŭnga" [Introducing occupations for women], *Pyŏlgŏngon* 5 (1927): 101.

76 Pak Yunjae, *Han'guk kŭndae.*

77 On missionaries' relations with the Japanese colonial medical administration and their mission boards, see Sonja Kim, "Missionaries and 'A Better Baby Movement' in Colonial Korea," in *Divine Domesticities: Christian Paradoxes in Asia and the Pacific,* ed. Margaret Jolly and Hyaeweol Choi, 57–83 (Canberra: Australian National University Press, 2014).

78 Yi Kkonme, "Ilbanin ŭi Hanŭhak insk kwa ŭiyak iyong" [A study on the general public understanding and utilization of traditional Korean medicine in the colonial period], in *Hanŭihak, singminji rŭl alt'a: singminji sigi Hanŭihak ŭi kŭndaehwa yŏngu* [The modernization of Korean traditional medicine during the colonial period], ed. Yonsei University Institute for History of Medicine, 137–154 (Seoul: Ak'anet, 2008).

79 "Saech'un ŭi sae hŭimang" [New hope for the new spring/youth], *Pyŏlgŏngon* 11 (1928): 31. Pak Insŏk's "Hwanjarosŏ ŭisa ego" [To doctors, from a patient] was the last vignette in the collection.

80 "Aktŏk ŭisa e ch'ŏlt'oe" [Crackdown on corrupt physicians], *Tonga ilbo,* November 11, 1933.

81 Chŏng Tŏkkŭn, "Hyŏndae ŭihak muyongnon" [The futility of contemporary medicine], *Tonggwang* 32 (1932): 88–90, 97.
82 "Hwangsŏl susŏl" [Hearsay, nonsense], *Tonga ilbo,* November 12, 1933.
83 "Sisang manhwa," *Pyŏlgŏngon* 33 (1930).
84 O Kŭngsŏn, "Ton pŏnŭn ŭisa poda pyŏng koch'inŭn ŭisaro," *Chogwang* 3 (1941), 204, cited in Pak Yunjae, "Singmiji sigi ŭisa kyech'ŭng ŭi sŏngjang kwa chŏngch'esŏng hyŏngsŏng," 150.

CHAPTER 3

Cloning National Pride

Science, Technology, and the Korean Dream of Joining the "Advanced World"

Inkyu Kang

The dawn of the new millennium saw a new set of buzzwords circulating among South Koreans, among them "cloning" and "biotechnology." This trend was indebted largely to Hwang Woo-suk, a veterinarian and professor at Seoul National University. He had risen to prominence in South Korea (Korea) after announcing that he had cloned a cow in 1999. Five years later, Hwang was elevated to a national hero after he declared in 2004 that he had succeeded in cloning human embryos using the somatic cell nuclear transfer method.

The following year, the star scientist stunned the world by proclaiming another triumph, which was even more splendid. His team, he said, had produced unique stem cell lines, for the first time in history, from skin cells of eleven people. Because stem cells have the potential to develop into a wide variety of cells that compose organs, bones, and tissues of the body, Hwang's claimed work meant a gigantic leap toward the creation of patient-specific body parts for transplant or cure. Most Koreans were ecstatic about the homegrown scientist's historic achievements. His nickname—Pride of Korea—sounded well deserved.

The government promised, at the drop of a hat, "as much financial support as Dr. Hwang needs." It had already invested heartily, spending more than ₩26 billion (about $24 million) to assist his research in 2005 alone as well as giving him the title of Supreme Scientist.[1] The previous year, the Roh

Moo-hyun administration awarded the highest honor—the Changjo (creation) Medal—for his outstanding accomplishments in advancing Korea's science and technology.

"Many of us were feeling worried, thinking that Korea is hopeless," then President Roh started his congratulatory speech to a rather gloomy tone. "Quite a few doubted Korea could be world-class." He concluded, however, on a more triumphant note: "You have planted confidence in us." At that moment, Hwang and his team were not just highly accomplished scientists; they were the embodiment of their country's hope and future—until everything came crashing down.

Investigative reporting by a Korean major network revealed that Hwang and his colleagues not only fabricated evidence but also committed serious legal and ethical violations in acquiring ova. The nation was devastated. Some of Hwang's avid supporters were in denial, but what swept over the country was a massive wave of humiliation. Because he was the pride of Korea, his fall was the shame of Korea. People asked the same question over and over: "How did this all happen?"

This chapter revisits the Hwang incident by critically reviewing the existing literature, putting it in social, political, and historical context. Several related studies have been published since the 2005 fiasco but one important question remains unanswered: Why did the worst scientific fraud in Korea's history occur during Roh's administration, which clearly had a democratic and liberal inclination? Some account for the scandal in terms of "blind patriotism" or "a culture of secrecy."[2] Confucianism or "East Asian culture" is routinely blamed as the usual suspect in the discussion of not only of Hwang, but also of wide ranges of ethical violations that Asian scientists commit. Many of these approaches are based on a monolithic, homogenous, static, and deterministic—if not downright Orientalist—notion of culture.[3] One serious pitfall these approaches tend to entail is turning a blind eye to historical contexts and social dynamics.

Tradition has been an easy target, too. Many studies point a finger at the practice of treating technoscience as a tool for industrialization since the mid-twentieth century. In doing so, a number of researchers find their answers from Park Chung-hee's military junta (1961–1979), or even from the Chosŏn dynasty, which ruled for five hundred years until it fell in the late nineteenth century. " 'Traditions' which appear or claim to be old are often quite recent in origin and sometimes invented." Without referring to this radical notion of tradition as an invention suggested by Eric

Hobsbawm, "tradition," which is far from being unchanging, cannot be used to justify an ahistorical claim that lacks the necessary context, nuance, and subtlety.[4]

In any country, tradition is a battlefield where different social and discursive forces collide. As Phillan Joung and Marion Eggert rightly put it, "Not only is 'tradition' necessarily at a loss for answers to these contemporary problems; in Korea, with its complex cultural heritage, 'tradition' itself has long been a field of contestation."[5] A vast portion of the literature on Hwang's misconduct and the role of the government centers around Korea's hastened industrialization or "compressed growth" (*apch'uk sŏngjang*). Interestingly enough, President Roh actually declared a break from the legacy of compressed growth in his speech at the National Assembly in October 2004. It was roughly a year before PD Notebook, an investigative journalism program, started gathering information about Hwang's ethical violations and the validity of his *Science* papers.

The president's statement, of course, does not guarantee that his government completely turned away from what its predecessors had done for years. As discussed later, Roh did rely on the rhetoric of *kwahak ipkuk* (nation-building through science), which was coined by Park, the army general turned president who ruled Korea for more than fifteen years. Some twenty-five years after the strongman's assassination, however, the young president, a former human rights lawyer, made it clear that the traditional state-run development plans had had their day.

Traditions offer continuity, but they also cause ruptures. This chapter sheds light on the latter, which existing studies have failed to capture. In a nutshell, it seeks to show a more nuanced picture of the Hwang case in its complexity and diversity.

Contextualizing the Tradition

"I felt electrified," said then President Roh in astonishment in Hwang's lab in December 2003. The veterinarian in the spotlight was offering a demonstration of cloning to the fifty-seven-year-old leader visiting him on the Seoul National University (SNU) campus. "This is not technology; it's magic." He continued, "I surely saw the possibility and hope for the era of a $20,000 per capita income."

Even before Roh took office, the government had already provided Hwang with sizable research funds since his cow cloning in 1999. After the dramatic lab meeting, even heftier investments poured in from both public and private sectors. It is estimated that Hwang and his team received more than ₩10 billion (about $85 million) for the four years from 2001 to 2005. Hwang was also offered the opportunity to form the World Stem Cell Hub, which opened on October 19, 2005, just before he was revealed to have used fabricated data (eventually he was charged with embezzlement and bioethics law violations). The name of the research institute seems to show many Koreans' long-dreamed dream along with its catchphrase: "Hope of the World, Dream of Korea."

A few months before the president's visit to Seoul National University, his administration announced the National Basic Plan for Science and Technology, proclaiming its vision of "realizing the second *kwahak kisul ipkuk* by establishing a science- and technology-centered society." The policy report lists three major industry areas: information and communication technology (ICT), bioengineering, and environmental technology.

The report also states its rationale: "Korea is turning into a knowledge-based society, where knowledge, information, and science technology are the key source for wealth and growth." The second point makes clear why the Roh administration placed unestablished, nonmanufacturing sectors like biotechnology front and center: "The high-added-value sector has shifted from manufacturing physical products to creating nonphysical scientific technology, knowledge, information and services."[6]

Roh also adopted the slogan "*kisul ipkuk,*" taking a highly critical stance against Park's rule. Oh Myung, the minister of science and technology Roh appointed, delivered a keynote speech in 2005 for the thirty-ninth anniversary of the Korea Institute of Science and Technology, which was also founded during Park's rule. Oh said that "the government is seeking to firmly establish the basis of the second *kisul ipkuk*" so that "it meets the needs of the times."[7]

At a glance, everything looked familiar: a new president is elected, and his or her administration proclaims "strategic industries." The president gives speeches on the significance of science and technology, highlighting how they can help increase the GDP growth rate and bring Korea one step closer to becoming a "developed nation." Officials release policy reports and start massive investments in the "new growth engine."

Many studies of Hwang's misconduct base their claims on this kind of one-dimensional, never-changing image of society. Is it plausible to assume that the two administrations, separated by a generation, followed the same tradition, and that Koreans were blinded by the identical nationalist agendas? What sets them apart is not just time. When Park, arguably the archetype of technoscientific leadership, declared Science Day in 1967, Korea was one of the world's poorest countries, its GDP per capita barely $161.[8] The level was half that of Senegal, and a significant portion of it came from foreign aid. By the turn of the twenty-first century, the country had risen to become the world's twelfth largest economy, after fighting off the second junta as early as 1987.

During the period between Park and Roh, Korea underwent important transformations, including in its major industries, economic structure, political system, and living standards. Further, international politics related to the Korean peninsula went through a sea change. This included the dissolution of the Soviet Union, the rapid decline of the North Korean economy, and normalized relations between South Korea and China, which was emerging as an economic powerhouse. In addition, the Asian financial crisis that hit Korea hard in 1997 left a permanent dent on every inch of the country.

After Hwang's fabrication was revealed, the *New York Times* reported that the Korean government "tried the same trick with biotechnology" it had used to "churn out cars and computer chips."[9] The Roh administration undoubtedly deserves its share of blame, but what the *Times* wrote does not seem to be an accurate evaluation. If the government's hardware mindset was at fault, why did it get into biotechnology in the first place when it could have simply kept churning out cars and computer chips?

Although encouraged by the renowned bioscientist's spectacular breakthrough, the government decision to plunge into uncharted waters looked abrupt, even odd, considering Korea's robust manufacturing sector. In fact, they did noticeably better throughout Roh's presidency (2003–2008) than any other period. Korean conglomerates, dubbed *chaebŏl,* were doggedly infiltrating global markets and setting new records in revenue and market share every year.

The Hyundai Motor Company celebrated the end of the first year of the new century with a record profit for 2001 of ₩1.17 trillion ($888.9 million), an increase of 75 percent from the previous year.[10] The automaker's net profit almost tripled to ₩2.96 trillion ($2.57 billion) by 2009, when *Time* magazine praised it as "America's most successful car company."[11] Samsung Electronics did an even more impressive job. Before becoming the world's top smartphone maker in 2011, the semiconductor giant dethroned

Sony as the biggest TV maker in 2006 and took the lion's share in the US cell phone market in 2008. Other manufacturing companies, such as LG and POSCO, also looked unstoppable.

Why, then, did the government make such a sudden, risky choice to jump into biotechnology? What socioeconomic conditions prompted the Roh administration to bet big on the untested and ethically controversial field? What implications does the Hwang debacle have in terms of Korea's long pursuit of science and technology that constitute the core of the country's developmentalist ideology?

Specificity and Universality

Science and technology have been emphasized since Park Chung-hee's military junta established the Third Republic in the early 1960s. Although their significance has been stressed as the inevitable method of escape from destitution and misery, science and technology have meant more than practical tools to improve living standards; they have defined Koreans' national vision, pride, and even identity.

During the two back-to-back military regimes, virtually any scientific or technological area was welcomed as long as it was thought to promise a positive economic impact, but the key strategic focus varied depending on the leader's economic and political agenda. The strategic choice was light industry for manufacturing clothes and shoes during the 1960s; heavy industry composed of steel, automobile, and shipbuilding in the 1970s; semiconductor and electronics in the 1980s; and information and telecommunication devices and cultural products in the 1990s.

Entering the 2000s, the government made a surprise move of picking biotechnology as one of the country's top new growth engines. This was a radical shift from a decades-long tradition. From the First Five-Year Economic Development Plan of 1962 to 1966 until the turn to the twenty-first century, virtually all key strategic sectors had involved manufacturing—from human hair wigs to NAND-type flash memory chips. Further, as discussed, Korea's vibrant manufacturing showed little sign of slowing down.

Like most presidents since Park, Roh relied on not only the old-fashioned nation-building catchphrase but also advancing science and technology. During his tenure, biotech industry started as an "offshoot of South Korea's 'Highly Advanced Nation' development agenda" in an attempt "to make South Korea into one of the seven leading biotech nations by 2007."[12]

A noteworthy difference, however, sets Roh's policy apart from that of his predecessors—the emphasis on basic scientific research. He was the first president in Korean history to choose developing fundamental science research as a key policy objective for his administration.[13] Most presidents before him prioritized applied research suitable for real-world application in order to achieve tangible results before leaving office, given that since 1987 Korea's presidential term has been five years with no possibility of reelection.

Roh's long-term approach to science and technology, including bioscience, should not be dismissed for the pitiful results. He believed that Korea would not be able to sustain its growth without radically transforming its economic system, which to date had capitalized on unequal growth. In his 2004 speech at the National Assembly, Roh asserted the need to reform the economic development model as well as his commitment to extending investment in science and technology. He contended that "the traditional measure [had] hit the wall," arguing for "a mid- to long-term strategy to fundamentally overcome the difficulties and problems our economy is facing."[14]

In February 2006, when Hwang fell flat on his face, the presidential Blue House announced a similar statement: "Compressed Growth, the Myth Is Over." It suggested that compressed growth and polarization in wealth are "twins whom the strategy of unbalanced growth gave birth to." It continued, "Only the bright side of the tactic was visible until the Asian financial crisis, but its drawbacks have become clear since the economic meltdown."[15]

A 2003 policy paper asserted the need to invest in bioscience and outlined the vision of the Roh administration: "Developing technology toward a healthy, life-valuing society." What the ambitious president said to Hwang in his lab—"I felt electrified"—was harshly criticized and ridiculed later, but his less-known remark on the spot supports the vision. "What you are doing now is extremely important and precious for human beings, regardless of its monetary benefit." He was applauded by adding, "As the president, it was a blessed privilege to eyewitness this deeply valuable endeavor."

Although Roh's plan derailed, the Hwang disaster erupted as Korea was emerging from the hardware mentality that had characterized the country for decades. Roh, despite being a political maverick, failed to challenge the myth of presidential leadership in science and technology. Why does a political leader have to stand as a beacon of light on the forefront of scientific innovations? Is it even possible for a scientific layperson to spearhead an unestablished territory that can be barely understood by very few experts?

One of the dangers of assuming this near-impossible leadership role is reliance on an exclusive group of prominent scientists. Even if the members happen to be well-willed individuals, would it be democratic decision-making to implement large-scale investment plans for society based solely on the opinions of a handful of scientists, especially when they are to be major beneficiaries of such decisions? Although Roh obviously saw the limitations of unequal growth, he ironically ended up gambling hugely on an elite group led by the charismatic veterinarian who had wowed the world.

Hwang Woo-suk looked spotless on paper. He was already a tenured professor at prestigious Seoul National University and had an impressive record of publications, including papers in both *Science* and *Nature,* two of the most reputable scientific journals globally. Well known for their rigorous peer-review standards, the journals shielded Hwang from growing suspicions to the end. Many of his supporters suspended doubt, believing that the journals would have retracted his papers had they been faulty. *Science* did not retract the flawed papers until SNU released the final report on the research allegations confirming that Hwang's 2004 and 2005 *Science* papers had been fabricated. The retraction decision was made almost two months after the first television reporting of the whistleblowing by the researchers who had worked with Hwang.

Even before SNU set up an internal investigation committee to launch a probe, quite a few Korean scientists had raised suspicions about the integrity of the research, and some of them even contacted *Science* directly. Meanwhile, a new controversy related to the stem cell images included in the 2005 game-changing article began to brew. A considerable portion of them were suspected to be erroneous or manipulated. "There is no reason to believe at the moment that it is a problem that affects the scientific outcome of the paper," said *Science* Editor-in-Chief Donald Kennedy concerning the additional dispute. This announcement ended up prolonging the academic life of the falling hero.

A large number of academic misbehaviors was disclosed around the world before Hwang's fall, but the selection standards or review procedures of scientific journals have not arisen as major issues in the cloning fiasco. Some scholars appeared to quarantine the case to within the boundary of Korea despite Kennedy's announcing that his journal "would tighten its review procedures and would work with *Nature* and other leading journals to try to identify fraudulent papers."[16] Herbert Gottweis and Robert Triendl, for example, argue that "responsibility for the Hwang scandal lies not with the peer-review system but with bad science governance."[17]

As in many other countries, highly selective journals like *Nature* and *Science* have come to bear almost absolute authority in Korea given the growing competition in academia. Some research universities have even begun paying cash incentives to faculty who publish papers in renowned journals indexed in the Science Citation Index, especially the "NSC journals" (*Nature, Science,* and *Cell*). The monetary rewards vary, but wealthy private universities such as Yonsei, Ewha, and Sungkyunkwan are known to have paid between ₩10 million and ₩30 million ($9,000 to $27,000) per publication in a prestigious venue, from the mid-2000s to the late 2010s. Why pay such generous incentives? The government routinely ranks universities to determine how much financial support to offer, and the number of faculty publications, especially in journals with impressive impact factors, is amply rewarded in evaluating a school.

This practice of giving heavy weight to internationally acclaimed journals, not to mention awarding financial compensation, was hardly part of a time-honored tradition of Korea. It was instead a result of the radical education restructuring initiated by the Kim Young-sam administration (1993–1998), the first civilian government in the South. In particular, the 5/31 Education Reform has had a profound impact since it was proposed in 1995, a decade ahead of the Hwang debacle. The policy proposal argued for the need to "restructure the entire education system in preparation for a knowledge-based society." It outlined forty-eight objectives, including diversifying K–12 school choice, reinforcing gifted and talented education, and reforming college admissions, but the main focus was on deregulation, decentralization, and competition among schools and teachers.[18]

Especially notable was the introduction of unequal, performance-based compensation for teachers and government funding for schools. Many of the proposed reforms were implemented during the succeeding administrations, including those of Kim Dae-jung and Roh Moo-hyun, turning the exclusive scientific journals into household names. Given that the newly enacted neoliberal education reform provided fertile soil for academic misconduct, the Hwang affair can be inquired into more accurately in terms of universality rather than specificity. In other words, the cloning fraud is related more closely to general neoliberalist tendencies than to Korea's unique tradition or culture.

Rupture and Continuity

The competition-oriented education system is becoming the norm in many parts of the world, including Europe, North America, the Middle East, and Asia. The controversial 1995 legislation of Korea was modeled after the Education Reform Act of 1988 of the United Kingdom, which had been criticized for making schools "compete for customers and their business expands according to their perceived success, determined mainly by test and exam results published centrally."[19] This Thatcherite agenda has inspired similar education reforms around the world, though the idea is rooted in the paradox of decentralization through centralization: the central government exposes education to market forces while tightening its grip on schools by tying funding to standardized test scores.

One of the results of the implementation of this measure-and-punish policy is widespread, organized cheating, which schools either ignore or even encourage. For example, it is often suggested that the Elementary and Secondary Education Act (ESEA) of the United States, commonly known as No Child Left Behind, has "led to some schools resorting to cheating, so as to increase test scores." Because federal funding is a lifeline the majority of schools need to stay afloat, teachers, principals, and even local governments are tempted to pull out all the stops for survival. Paul Thomas criticized this evidence-based education policy in his *Newsweek* article "No Child Behind Is a Cheats' Charter."[20]

Although ESEA was passed during Lyndon B. Johnson's presidency in 1965, it was reauthorized by George W. Bush in 2001 and then replaced with Every Student Succeeds Act under Barack Obama. When an Atlanta jury found eleven teachers and administrators guilty of racketeering and other crimes in 2015 in a massive wave of test cheating in almost forty states and Washington, DC, the *Washington Post* pointed to the decades-long education policy.

> No Child Left Behind, President George W. Bush's chief education initiative, and then Race to the Top, President Obama's central education program, placed increasingly high stakes on standardized test scores. They had to go up, or else there would be negative consequences not just for students but schools and teachers and principals. Such test-

> ing mandates were coupled with a "no excuse" management push by school reformers who said teachers had, well, no excuse not to raise their students' test scores.[21]

This trend is not unique to the United States, obviously. Comparable cases have been reported from other countries; connections have even been suggested between the norms of free-market capitalism and both economic and academic frauds—such as the direct relation proved between neoliberal values of power and personal success and the condoning of cheating.[22]

It would thus not be productive to link the "worst of a wave of test cheating" to "American tradition" without proper contextualization. It would not be fair, by the same token, to equate Obama with Bush or Johnson for the similar tendencies in each of their education policies. The same can be said of Roh and Park, arguing for carefully considering both similarities and differences. Taking off the straitjacket of tradition is a good starting point for dissecting the cloning fraud to paint a more salutary picture. Roh's presidency clearly marked a rupture, but it also revealed continuity: technological utopianism.

In February 2005, Korea Post, the national postal service, issued a commemorative stamp in honor of Hwang's cloning of human embryos in the previous year. Although this came before Hwang's declaration of an even more dramatic achievement in May 2005, the stamp design provides a glimpse of how Hwang's claimed breakthroughs articulated with existing discourses of science and modernity. The stamp was unusual in many aspects. First, it was one of the rare cases in which a recent achievement by a living scientist was celebrated on a Korean postage stamp. The design was unique as well.

The horizontally long—very long—stamp has two parts: the left half is a montage of four pictures showing the process of nuclear transfer; the significantly larger right half features a large picture of a stem cell. Against this two-tier backdrop sits a time-lapse image of a man's white silhouette. On the right border of the left section, he sits in a wheelchair. Entering the right section, he rises, walks, and leaps. The continuous images end dramatically with a man embracing his beloved, who is wearing a skirt.

The two sections are divided by their use of the colors red and purple, but connected by the silhouetted man who "progresses" across the border. On the left is a series of images of injecting a nucleus into an egg cell, on the right a successfully cloned stem cell, big and bold. The stem cell, in pink, lies at the center. Because the background is dark monochromatic purple, the light pink tone shines like a flower in full bloom or the sun behind the couple holding each other. The right side of the stamp appears to symbolize a rosy future that awaits the nation—or the world. Eunjung Kim offers an interesting interpretation of this design:

> This process captures how curative science enables intimate relations, making visible the assumption that normative functioning is the precondition of social inclusion. . . . As if filmed in stop-motion, the five figures appear as a single body being fast-forwarded through time. Adding to this effect of animation, a hologram creates depth and glow in the horizontal movement of "progress" to arrive at the privileged, nondisabled body.[23]

Kim's analysis of the images at the individual level—"normative functioning" as "the precondition of social inclusion"—can be extended to international relations. A possible interpretation would be that the underdeveloped state, represented by the left half, is seen as a form of disability that keeps the country from being included and recognized in the international community as a competent nation. It is not necessarily because Korea is plagued by "deep-rooted insecurities," as the *New York Times* claimed; it can be a reasonable desire to play an appropriate role on the global stage.[24]

During the rule of Park Chung-hee, science and technology were established as symbols of modernity and industrialization. Park emphasized technoscience as a way to revive the postwar economy by following in the footstep of Japan's post–World War II technocracy. He had other motives, though, beyond boosting economic growth. One was justifying his unelected, unchosen political power: no better way to give people a tangible sense of progress than to show them images of asphalt paved highways, multistory buildings, shiny chemical plants, or orange-colored molten metal pouring from a blast furnace in a steel mill, calling them "science and technology."

Technoscience, focusing on the practical and immediate applications of science and technology, was also essential in responding to potential threats from North Korea, which was wealthier and better armed until the end of the 1970s. Park sought legitimacy in staunch anticommunism, although he had once been sentenced to death for participating in a communist cell, feeling the North as a threat not only for the South but also for his own safety.[25] A few attempts were made to assassinate Park, including the 1968 incident when thirty-one North Korean commandos tried to attack the Blue House. Kim Sin-jo, one of the two surviving team members of the raid, blurted out to a reporter asking what his mission was, "I'm here to cut the throat of Park Chung-hee."

Since South Korea achieved democracy in the late 1980s, the motivation behind the country's technoscientific push has changed. The associations of science and technology with national security and the heavy and chemical industry drive have weakened dramatically, replaced by new industries and different symbolic meanings. The civilian presidents since the early 1990s strove to differentiate their administrations from their authoritarian predecessors. Kim Young-sam's radical education reform was along this line, and Roh's vision for biotechnology was set in a similar vein in a growing neoliberal atmosphere.

The Intersection of Developmentalism and Neoliberalism

Park Chung-hee (1917–1979) and his archrival Kim Il-sung in the North (1912–1994) are arguably the most controversial politicians in Korea's modern history. After becoming president through the 1961 coup, Park did anything he could do to remain in power—including amending the constitution, dissolving the National Assembly, declaring a state of emergency, arresting journalists and shutting down newspapers, and executing his critics for "acts of treason." After sixteen years of iron-fisted rule, Park met death while seeking lifelong presidency—shot dead by his own security chief at a dinner party.

Park's rule was oppressive and cruel by any measure. Even his supporters admit that he was a ruthless dictator. Such a consensus, however, shows only part of the picture. Quite a few Koreans have a strong nostalgia about the strongman. "He was a tyrant, but he dug us out of poverty." This statement effectively summarizes the attitude of Park's admirers, most of whom are old enough to have experienced the Korean War (1950–1953) and the severe famine that followed. Their evaluation of Park may show more about their traumatic experience than about their political consciousness, but the so-called Industrialization Generation has never hesitated to give Park the credit for Korea's rapid economic growth. At the center of his economic policy for "guided capitalism" were science and technology.

President Park Chung-hee's rule was characterized by the propaganda of *ipkuk* (nation-building). Three strategies were proposed to build the nation through export (*such'ul ipkuk*), science (*kwahak ipkuk*), and technology (*kisul ipkuk*).[26] Key scientific research and education institutes such as the Korea Institute of Science and Technology and the Korean Advanced Institute of Science were established in 1966 and 1971, respectively, during which time Park was at his peak in power.

Technological utopianism found in the country can be explained partly by the thirty-five years of Japanese rule (1910–1945), which most Koreans find shameful and traumatic. Under colonial rule, Koreans were forced to adopt Japanese-style names, and tens of thousands of young women were drafted as sex slaves for Japanese soldiers during World War II. Although Korea gained independence when the Empire of Japan surrendered to the Allies, now Koreans had to see their country divided by foreign superpowers. The division led to the Korean War, causing total deaths of more than 1.2 million.[27] Many Koreans believe these catastrophic experiences could have been avoided had Korea modernized itself sooner.[28]

It is interesting to see how two recent presidents, Lee Myung-bak (2008–2013) and Park Geun-hye (2013–2017), won their elections by making the old-fashioned promise to "transform Korea into a developed country"—over a decade after the country joined the Organisation for Economic Co-operation and Development in 1996, which is often referred to as the "rich-country club." The phrase "developed country" (*sŏnjin-kuk*) has never failed to fascinate Korean voters since the 1960s. Korea became the twelfth largest economy in the world by 2007, its per capita income at more than $20,000 that year and more than $23,000 in 2012. Voters, however, still seemed eager to see their country join the "advanced world," which is a mirage that keeps receding as they advance.

Science and technology have long been identified with economic development, or national income to be precise. New scientific inventions or discoveries are routinely evaluated in terms of economic benefits they can possibly bring. Technology is often considered "a means to earn parity with developed Western societies," and "compet[ing] head on with the West" is seen as "the ultimate barometer of scientific and technological progress."[29] When Dr. Hwang made world headlines in 2005 that he had successfully cloned patient-specific stem cell lines, the South Korean government and many news media declared in unison that the success would be worth ₩33 trillion ($30 billion).[30] They also claimed that his scientific achievement would help Korea leap into "the league of advanced nations."[31]

One of the first things Koreans learn in school about their country is that Korea is a "small country with few resources" and "a shrimp among whales." This commonsensical belief about Korea's geopolitical situation has led its government to actively pursue a globalization policy, especially after the 1997 financial crisis. President Roh was an avid believer of the "sandwich theory," arguing that Korea is being squeezed between hi-tech Japan and emerging China. Roh thought biotechnology would provide a breakthrough, as his predecessor Kim Dae-jung led the New Economy by investing heavily in ICT.

Because Roh witnessed the rapid emergence of Chinese manufacturers like Lenovo—it shocked the world by announcing the acquisition of IBM's PC division in 2004—he came to believe that short-term tactics relying on technoscience cannot put Korea ahead of other players for very long. This realization led him to focus on basic science and technology and unestablished fields, such as bioscience, into which most rivals were not daring to jump. A speech Roh gave at the National Assembly reveals his state of mind:

> We need to pursue an innovation-led economy to increase our growth potential by improving the efficiency of the overall industries. We do not have enough time. Developing countries like China are catching up to us at a faster rate than the pace at which we are narrowing the technology gap with advanced nations.

It thus can be said that the cloning fraud took place at the intersection between Korea's developmentalist legacy and neoliberal pressures. Roh did not hide his neoliberal orientation in saying, "Now power has been handed to the market [from politics]."

This trend continues. As soon as Moon Jae-in, Roh's political heir, was elected to the presidency in 2017, he officially announced the slogan of the Fourth Industrial Revolution as part of a major economic plan. It came with deregulatory policies that included loosening privacy protections for personal data. Such measures, Moon said, were "to encourage scientific and technological innovations to catch the opportunity to transform the Korean economy from follower to leader."

Nationalism, Culture, and the Mass Media

Hwang Woo-suk did know how to stir up patriotic fervor. On his way back from a conference in the United States, for example, he stated, "I hoisted the national flag on the hill of U.S. bioengineering technology."[32] He also asserted in an interview, "Science may not have nationality, but a scientist does." He even announced a project to clone Siberian tigers, the almost extinct but nonetheless sacredly meaningful animal in Korea, as if he were on a mission to realize his fellow Koreans' wish to roar into the world's top spot.

It would be fair to say that quite a few Koreans stood up against the negative reporting on Hwang, and that some reacted out of a strong sense of patriotism, even after his legal and ethical violations became clear. Hwang's supporters' impassioned protests took the investigative television program that had questioned the fidelity of his research off the air for two weeks. Before the show was temporarily canceled, its twelve regular advertisers had to stop sponsoring the program amid angry accusations and threats of boycotts.

Meanwhile, an online fan club ("I Love Hwang Woo-suk") was quickly gathering members to support their beloved scientist as he went down in flames.[33] The national anthem greeted visitors as background music. Such an appeal for patriotism may not have been surprising considering Hwang's status of national hero, or even national interest itself to some. It was not everything. Something far more disturbing was happening; the fan site started a campaign for online ova donations so that Hwang could continue his stem cell research.

The main page of the site was covered with more than seven hundred flowers in barely two weeks. They were Roses of Sharon, South Korea's national flower. Each flower image symbolized a woman who had pledged to donate her eggs. The *Washington Post* wrote that Koreans were "blinded to truth," and "virtually nobody believed" when "Korean investigative television reporters . . . exposed massive fraud and ethical breaches by Hwang."[34] The truth is, however, that the scientist was thrown into a hot debate as soon as the first exposé was aired. Those committing to ova donations plunged to almost nil within a month, and the club manager soon announced the site would no longer publish the number of pledges.[35]

Not all supporters were misguided or blinded by national pride, either. Some were intrigued by Hwang's personal narrative as delivered routinely by the mass media. A boy from a poor family had worked his way earnestly through all the hardships to eventual success. Hwang enjoyed talking about his grueling work schedule without a weekend, "Mon, Tue, Wed, Thurs, Fri, Fri, and Fri. . . ."[36] One might say his diligence better represents the Protestant work ethic than the Confucian counterpart; the scientist used to be a devoted Christian.

Furthermore, nationalism did not work only to cover up or keep silent about Hwang's fraud and misconduct. Some of those who participated in exposing what Hwang did believe that not allowing such unethical academic behavior is really loving their country. A former colleague of Hwang went on to expose to the PD Notebook reporters untold truths about the disgraced clone expert's questionable research activities. The whistleblower later confided that he asked the team, "What do you want? National interest or the truth?" Without missing a beat, the head producer replied, "The truth is national interest."

It is easy to hastily label others as nationalistic. As Terry Eagleton suggests, ideology is like bad breath—a problem only others have. It would also be simplistic to find fault with Korean culture, Confucianism, or Asian culture for what Hwang did. Would it be valid to blame the fraud case of Elizabeth Holmes, founder and former CEO of Theranos (2003–2018), who falsely claimed that her company had devised a revolutionary blood testing, on American culture, Christianity, or Western culture?

Young Hee Lee confronts the mainstream view that the Hwang affair was a "Korea-specific phenomenon."[37] Common factors regularly attributed to the incident include culture, nationalism, inferiority complex, corruption, state-guided capitalism, and greed—which do not appear to be limited to Korea even if they were all relevant. The claimed Korean-ness of the scandal, however, loses its pertinence as soon as comparable, if not worse, cases from other countries are pointed out.

Lee admits the "Korean specificity" of the fact that the Hwang followers did not withdraw their sympathy even after the veterinarian confessed to the use of fabricated data. Considering the near unanimous support from the mainstream media, however, it is difficult to blame people who had very limited and biased information on the unraveling situation. The Hwang scandal was a multilayered problem caused by a wide range of structural and accidental factors, but the mass media was definitely a major player from the beginning to the end.[38]

In 1997, an interesting poll was published in several newspapers—when Hwang was about to get out of obscurity. "Who would you like to clone?" The largest proportion of respondents (15.7 percent) among those who named a person voted for Park Chung-hee. Ranked second was Kim Koo (1876–1949), a leader of Korea's independence movement and leading reunification activist during the early and mid-twentieth century (9.3 percent). Sejong the Great (1418–1450) came in third place (6.5 percent); as the creator of the Korean alphabet, he is the most respected of all kings.

The survey results might make some wince, because the first cloning candidate—although imaginary—was a ruthless dictator. A closer look at the poll data, however, may be relieving to them. Almost twice as many people who picked Park answered "Nobody" (35.5 percent), and those who said "I don't know" accounted for a sizable 22.8 percent. In other words, almost 60 percent of Koreans found the idea of cloning a human unpalatable. Such a negative attitude was to change dramatically in less than a decade, as the Hwang case proved. What changed their mind so quickly?

Until 1999, when Hwang made his first big break by cloning a cow, most news media in Korea were strongly opposed to genetic engineering. In 1993, for example, the *Seoul Shinmun* ran an op-ed column titled "There are things humans dare not to do." In the same year, the *Segye Ilbo* published a critical piece on genetic modification of living creatures. The article concludes, "It is scary that such genetic experiments might end up bringing a world of Frankenstein-like monsters." The *Hankook Ilbo,* a relatively liberal newspaper, also expressed similar worries in its 1997 editorial "Fear of cloning life." The writer was concerned about the influence of Dolly, the first mammal cloned from an adult somatic cell in the United Kingdom.

The dystopian view of biotechnology of the Korean press shifted notably through two monumental stages: Hwang's cloning of a Korean cow in 1999, and his claimed creation of human embryos through cloning in 2004. The first achievement softened the fear of animal cloning, but the media still remained hostile about cloning human cells until Hwang's team published the groundbreaking research in *Science.* As the press changed its position, public opinion went through a similar transformation. In this sense, the Hwang incident was a media event from the beginning. The mass media put him on a pedestal and turned a bad case worse with their uncritical reporting when doubts were raised by citizens.

The media in Korea were under the direct influence of the government during the successive military regimes from the early 1960s to the late 1980s. For almost four decades, the media were mobilized for "national development," and the government even provided reporting guides on specific events and issues for broadcasting networks and newspapers.[39] However, Korean media have quickly become incorporated into the market system since the early 1990s. The unbalanced and inaccurate reporting that stirred Hwang's supporters was more serious among commercial media than it was among their public counterparts. It was indeed a public television network, MBC, that broadcast the first exposé. Lack of diversity in the media is an issue shared by many countries, yet another reason to explain the Hwang case in terms of universality rather than typicality.

Conclusion: The Cloning Scandal as a Site of Struggle

Since Park took office in a coup d'état, science and technology have been promoted as the driving force for industrialization with the help of the mass media. They were not just a way to achieve economic growth or to gain a dominant market position in global competition. To justify their unconstitutional power, the junta leaders had to paint a utopian picture of the consequences of their rule, and science and technology came in handy to boost hope and pride. During the process, science and technology have become the ultimate symbol of what the future will bring—from skyscrapers to dense highways as illustrated in government public relations materials during the 1970s. As time passed, those images have been updated to a space rocket piercing the sky and a paralyzed patient walking again.

Picking biotechnology was clearly along the line of choosing "strategic industries." It would be simplistic to argue, however, that the government's intention lay entirely in finding a cash flow and that the decision came out of its same old hardware mindset. It is important to remember why the Roh government attempted to diversify Korea's industries beyond its vigorous manufacturing. The 2003 policy paper that promoted biotechnology emphasized the vision of the government: "developing technology toward healthy, life-valuing society." Although the plan went sour, the Hwang incident broke out while Korea was trying to break the mold. It would be equally unfair to say Koreans supported Hwang out of blind nationalism; many of them had high hopes about the humanitarian use of stem cells as well. It was Koreans who revealed the truth—the whistleblower, journalists, researchers, and average people.

Korea's move toward a knowledge-based economy did not come out of nowhere. The country's geopolitical situation instilled its people with a competitive attitude, which has been strengthened since the 1997 Asian financial crisis. The neoliberal restructuring following the economic meltdown threw its public sector into the market economy, including health care and education. Increased academic dishonesty and misconduct is related to the added pressure of productivity characterized by the publish-or-perish mantra.

In December 2012, Park Geun-hye was elected president; she is the daughter of the country's first junta leader. Roughly two decades ago, Koreans chose the father as the most desirable specimen to clone. The daughter's victory seemed like a political realization of the cloning survey. Winning over half of the voters, Park's presidency appeared to prove that Korea was still on the path laid out by her father. As an evangelist of neoliberalism, she accelerated the privatization of education and public health services. What ensued was the dire socioeconomic state where many Koreans, especially the young generation, found themselves. They satirically referred to it as Hell Chosun.

The gloomy unofficial title of the country instantly became a popular buzzword, making it one of the words of the year selected by a major newspaper in 2015. Considering that Chosun is a Korean kingdom that tragically fell in the late nineteenth century, it would not be difficult to guess young Koreans' frustration from the naming. Many have even lost the will to live, which accounts for Koreans having one of the highest suicide rates among developed nations. It would be no surprise, then, that Korea's fertility is the world's lowest. Those who have given up on hope are unlikely to be hopeful about their children's future.

In 2016, President Park's approval rating plunged to a record low of 5 percent. It was revealed that the new president had relied on a small inner circle, and especially on her old friend, in making her public and private decisions. The unofficial aides allegedly twisted the arms of big businesses such as Samsung, CJ, KT, Lotte, and POSCO to get donations of millions of dollars. Such practices appalled even those feeling nostalgic about her father. As a result, hundreds of thousands protested in Gwanghwamun Square for weeks, calling for Park to step down.

Gwanghwamun Square was where a million citizens had gathered to end dictatorship, the dark legacy left by Park Chung-hee. Standing on the same spot, the largest number of protesters since 1987 asked for the resignation of his daughter. Ironically, it was also the exact spot where thousands of Hwang's admirers had gotten together in support of him a few years previously. Have Koreans finally found a new path now that it has been over a decade since the stem cell fraud? It is hard to say, because history is a multifarious—and often contradictory—process, full of ebbs and flows. One thing is clear, however: the Hwang scandal was a site of struggle where diverse norms, values, ideologies, and desires collided head on with each other—like Gwanghwamun Square itself.

Notes

1 Unless otherwise noted, the exchange value of the US dollar is calculated using the average exchange rates for the corresponding period.
2 Herbert Gottweis and Robert Triendl, "South Korean Policy Failure and the Hwang Debacle," *Nature Biotechnology* 24, no. 2 (2006): 141–143.
3 Leo Kim, "Explaining the Hwang Scandal: National Scientific Culture and Its Global Relevance," *Science as Culture* 17, no. 4 (2008): 397–415.
4 Eric Hobsbawm, "Introduction: Inventing Traditions," in *The Invention of Tradition,* ed. Eric Hobsbawm and Terence Ranger (Cambridge: Cambridge University Press, 1983), 1.
5 Phillan Joung and Marion Eggert, "The Cloning Debate in South Korea," in *Cross-Cultural Issues in Bioethics: The Example of Human Cloning,* ed. Heiner Roetz (New York: Rodopi B.V., 2006), 173.
6 NSTC, "National Basic Plan for Science and Technology," May 1, 2003, http://archives.knowhow.or.kr/m/policy/publication/view/19235?page=3.
7 Myung Oh, "Laying Ground for 'the 2nd Kisul Ipkuk,'" *Science Times,* February 25, 2005, http://www.sciencetimes.co.kr/article.do?todo=view&atidx=0000009244.
8 Kim Kŭnbae [Geun Bae Kim], "Pak Chŏnghŭi chŏngbu sigi kwahak kisulŭl ŏttŏk'e pol kŏs'inga? Kwahak taet'ongryŏng tamnonŭl nŏmŏsŏ" [How to understand science and technology of the Park Chung Hee regime: Beyond the discourse of the 'president of science'], *Yŏksa pip'yŏng* (Spring 2017): 143.
9 Choe Sang-Hun, "Lesson in South Korea: Stem Cells Aren't Cars or Chips," *New York Times,* January 11, 2006, https://www.nytimes.com/2006/01/11/science/lesson-in-south-korea-stem-cells-arent-cars-or-chips.html.
10 Don Kirk, "At Hyundai, Profit Surges to a Record," *New York Times,* February 7, 2002, http://www.nytimes.com/2002/02/07/business/worldbusiness/07iht-hyundai_ed3_.html.
11 Douglas A. McIntyre, "Hyundai: America's Most Successful Car Company," *Time,* September 3, 2009, http://content.time.com/time/business/article/0,8599,1882987,00.html; and Jane Han, "Hyundai Motor Profit Doubles," *Korea Times,* January 15, 2020, http://www.koreatimes.co.kr/www/news/biz/2016/02/123_59899.html.
12 Herbert Gottweis and Robert Triendl, "South Korean Policy Failure and the Hwang Debacle," *Nature Biotechnology* 24 (2006): 141–143.
13 Seong Deuk Ham and Da Sung Yang, "A Study of the Presidential Leadership for Science and Technology in Korea," *Korean Political Science Review* 46, no. 1 (2012): 141–173.

14 Roh Moo-hyun, "Che-250-hoe chŏnggi kukhoe sijŏng yŏnsŏl" [A speech given at the National Assembly in 2004], http://www.korea.kr/archive/speechView.do?newsId=132013160&pageIndex=280&srchType=title&srchKeyword=.

15 Blue House, *Policy Briefings*, "Apch'uk sŏngjang, kŭ sinhwanŭn kkŭnnatta" [Compressed growth, the myth is over], February 21, 2006, http://www.korea.kr/special/policyFocusView.do?newsId=135083871&pkgId=40000011.

16 Janice Hopkins Tanne, "*Science* Will Tighten Standards after Retracting Stem Cell Papers," *British Medical Journal* 333 (2006): 1189.

17 Gottweis and Triendl, "South Korean Policy Failure," 142.

18 Byungchang Sung, "The Education Policy of the Participatory Government," *Education Review* 12 (Summer 2003): 22–24.

19 Peter Wilby, "Margaret Thatcher's Education Legacy Is Still with Us—Driven on by Gove," *The Guardian*, April 15, 2013, https://www.theguardian.com/education/2013/apr/15/margaret-thatcher-education-legacy-gove.

20 Paul Thomas, "No Child Behind Is a Cheats' Charter," *Newsweek*, March 21, 2015, https://www.newsweek.com/no-child-left-behind-cheats-charter-315557.

21 Valerie Strauss, "How and Why Convicted Atlanta Teachers Cheated on Standardized Tests," *Washington Post*, April 1, 2015, https://www.washingtonpost.com/news/answer-sheet/wp/2015/04/01/how-and-why-convicted-atlanta-teachers-cheated-on-standardized-tests.

22 Caroline Pulfrey and Fabrizio Butera, "Why Neoliberal Values of Self-Enhancement Lead to Cheating in Higher Education: A Motivational Account," *Psychological Science* 24, no. 11 (2013): 2153–2162.

23 Eunjung Kim, *Curative Violence: Rehabilitating Disability, Gender, and Sexuality in Modern Korea* (Durham, NC: Duke University Press, 2017), 2.

24 Norimitsu Onishi, "In a Country That Craved Respect, Stem Cell Scientist Rode a Wave of Korean Pride," *New York Times*, January 22, 2006, http://www.nytimes.com/2006/01/22/science/in-a-country-that-craved-respect-stem-cell-scientist-rode-a-wave-of.html.

25 Yong-Sup Han, "The May Sixteenth Military Coup," in *The Park Chung Hee Era: The Transformation of South Korea*, ed. Byung-Kook Kim and Ezra F. Vogel (Cambridge, MA: Harvard University Press, 2011), 35–57.

26 Judith Cherry, *Korean Multinationals in Europe* (Surrey, UK: RoutledgeCurzon, 2000), 39.

27 Bethany Lacina and Nils P. Gleditsch, "A New Dataset of Battle Deaths," *European Journal of Population* 21 (2005): 154.

28 Sang-Hyun Kim, "Pak Chŏnghŭi chŏnggwŏn sigi chŏhang seryŏk ŭi sahoe kisulchŏk sangsang" [Sociotechnical imaginaries and dissident

groups during the Park Chung Hee regime], *Yŏksa pip'yŏng* (Fall 2017): 321.

29 Dennis Posadas, "The Dark Side of Asia's Technonationalism," *AsiaMedia,* July 24, 2006, http://www.international.ucla.edu/cnes/article/49701.

30 Heejin Koo and William Sim, "Fake Stem Cell Research Work May Cost Korean Industry Billions," Bloomberg, January 12, 2006.

31 Sŏ Kyŏngho, "10-tae sŏnjin'guk chin'ip chŏllyak [Strategies to get into the league of advanced nations]," *JoongAng Ilbo,* June 30, 2005, http://article.joins.com/article/article.asp?Total_ID=1626025.

32 B. J. Lee, "Cloning College," *Newsweek,* February 29, 2008, http://www.newsweek.com/id/53268?tid=relatedcl.

33 "I Love Hwang Woo-Suk!," http://cafe.daum.net/ilovehws.

34 Anthony Faiola, "Koreans 'Blinded' to Truth about Claims on Stem Cells," *Washington Post,* January 13, 2006, http://www.washingtonpost.com/wp-dyn/content/article/2006/01/12/AR2006011202404.html.

35 Tae-Ho Kim, "How Could a Scientist Become a National Celebrity? Nationalism and Hwang Woo-Suk Scandal," *East Asian Science, Technology and Society* 2, no. 1 (2008): 37.

36 Kim, "How Could a Scientist," 28.

37 Yi Yŏnghŭi [Young Hee Lee], "Hwang Usŏk sat'aenŭn ŏlmana Han'gukchŏk in'ga?" [How much Korea-specific phenomenon is Hwang's scandal? Reading universality and specificity of Hwang's scandal], *Kwahak kisulhak yŏnguk* [Journal of Science & Technology Studies] 7, no. 2 (2007): 31–42.

38 Jaeyung Park, Hyoungjoon Jeon, and Robert A. Logan, "The Korean Press and Hwang's Fraud," *Public Understanding of Science* 18, no. 6 (2009): 653–669.

39 Jukka Pekka Jouhki, "Korean Communication and Mass Media Research: Negotiating the West's Influence," *International Journal of Communication* 2, no. 1 (2008): 254.

PART II

RESTORING MINDS AND BODIES

CHAPTER 4

The Suicidal Person

The Medicalization and Gendering of Suicide in Colonial Korea

Theodore Jun Yoo

> "스토-ㅂ" [Stop], This is the last port. From time to time, ownerless hats hang on the railings fluttering in the wind bidding adieu. At the base of the pillars, large weeping bowls have been placed for young girls . . .
>
> —Kim Ki-rim, "The Han River Footbridge"

On July 27, 1927, the *Tonga ilbo* reported that a woman sporting a trendy *hisashi-gami* haircut had plunged off the Han River Bridge around 3:00 a.m. the previous day. A policeman from the Yongsan precinct, who was making his daily rounds, spotted a woman in the river and rescued her. The victim, later identified as Kim Sŏng-hŭi, was a student at Kyŏngsŏng Girls' High School and the wife of Cho Kŭn-sŏl. According to the police, Kim and Cho, both natives of North Hamgyŏng Province, had relocated to Seoul three years after they were married. Kim had demonstrated her resourcefulness and industriousness by enrolling at a vocational training center and later working at a fabric store. She had scrimped to save ₩400 to pay for her husband's high school education and also financed his trip to Japan so that he could complete his university studies in Tokyo. Just a month before her suicide attempt, Kim received the devastating diagnosis of myofascitis (acute muscle spasms), which led to a series of misfortunes. The husband whom she had supported for years demanded a divorce, claiming that he could no longer live with a sick person. As her illness progressively became worse and she lost her ability to earn a livelihood, a despondent Kim wrote a suicide note describing her cold-hearted husband and her miserable life. Unable to locate any family members, the Yongsan precinct detained Kim, placed her on a suicide watch, and offered to pay for her hospital bills.[1]

Was Kim a *mich'in yŏja* (crazy woman), or was her suicide simply part of a broader social phenomenon affecting a modernizing Korean society? Between 1910 and 1942, the Government-General of Korea reported a total of 54,053 suicides among Koreans.[2] Such alarming statistics, as well as sensational stories in the press, fostered a sense of crisis. Explanations were diverse, reflecting the competing agendas that shaped the discourse. On the one hand, the colonial authorities viewed the rise in suicide rates in Korea as an indicator that their project of modernization was working. Suicides were simply a by-product of the processes that Korea needed to undergo. That is why the colonial government documents rarely ever grappled with solutions to prevent suicides. The Korean press, on the other hand, treated suicides as emblematic of the tragic consequences of colonial modernity, which transformed Korea mores (especially in the area of marriage and family) and economic circumstances. At the same time, perhaps paradoxically, it also used suicide to deliver internal critiques of Korean society. In this sense, the press also gradually adopted and internalized the language of medicalization that the colonial state and medical experts used as a point of reference to explain the social problems of the day. Indeed, the pathologization of deviant behavior as a neurological disorder contributed to a broader discourse on suicide as a measure of social health, which placed people's lives under increasing scrutiny. This chapter examines these competing attitudes toward *chasal* (suicide) under Japanese colonial rule by analyzing traditional approaches and the reports of suicide in the Korean press.

Traditional Approaches to Suicide

In the 1920s, a new popular notion emerged that Korean society was being plagued by an "epidemic" of suicides. The Korean press blamed various facets of modern life—internal migration, rapid industrialization and urbanization, the end of a rigid class system, moral decadence, and other social stressors—as fostering a new type of pathological behavior that traditional remedies could not, or could no longer, control. This marked a shift from the traditional understanding of suicide in Korea, which had both a political and social meaning. In political terms, the word *chagyŏl* (individual suicide) signified an ultimate act of honor or atonement for a crime whereby the state granted a corrupt official, captured enemy officer, or a *panyŏk chwae-in* (enemy of the state) the option to kill oneself to avoid facing public humiliation. It was customary for the accused to declare loyalty to the king before drinking lethal poison.

In the *Chosŏn wangjo sillok* (The veritable records of the Chosŏn kings), some suicides were more social in nature. It mentions some 435 cases of suicide during the Chosŏn period (1392–1910), which allegedly resulted from a wide range of motives: facing embarrassing defeat in the hands of political opponents, unrequited love, *urhwa* (accumulated stress in social life), *pun* (feelings of anger due to failure), *ŏgul* (feelings of social unfairness), regret, and failure to live up to certain moral expectations.[3] Although official sources that explicitly examine individual cases among commoners are scarce, the *Hyŏngsa p'allyejip simrinok* (Criminal case studies, trial documents to the king), which were published between 1776 and 1800 during King Chŏngjo's reign, sheds light on possible motives for suicide among commoners in the late Chosŏn period. For example, in 1781, Pak Sŏngje, a commoner who lived in Yangyang, Kangwŏn Province, was falsely accused of stealing and endured a painful leg-screw torture. Overcome by shame and aggrieved by his inability to defend himself, he decided to hang himself. In another case in 1784, two neighbors were rumored to have fought over enlistment in the *sok-ogun,* a special unit for commoners. Yu Yŏn-in felt embarrassed and angry after quarreling with his neighbor and hung himself.[4] These two instances might not be representative of all suicides during the period, *punsa* (killing oneself because of indignation) was the most common reason why commoners killed themselves.[5] Given the paucity of sources, one can only surmise that honorable deaths or politically motivated suicides received attention in the public realm but the majority of suicides remained unreported or kept a family secret.

Yet another form of "sanctioned" suicides was the system of state canonization of widows (*yŏllyŏ-jŏn*), which promoted the Confucian view of morality and popularized the practice of widow suicide during the Chosŏn dynasty. The *Chosŏn wangjo sillok* mentions a total of 326 individual cases. Kim Chong-sŏng, however, suggests a much higher figure if one includes other official records. According to Kim, 694 faithful young widows followed their husbands to death.[6] Perhaps the most well-known mass suicides by widows took place during Toyotomi Hideyoshi's seven-year depredation of Chosŏn, when 356 women killed themselves to preserve their chastity and were later honored posthumously for their "virtuous acts."[7] Although these official sources might suggest that the imposition of a set of moral and social obligations (such as chastity ideology, three obediences, five human relationships) on society could have induced people to commit suicide, other triggers are also possible. For instance, folk culture, represented in *t'alch'um* (masked dance) and *p'ansori* (traditional epic song), stressed that unfair social situations, commonly precipitated by violence, discrimination, isolation, poverty, an overbearing mother-in-law, or infidelity could lead people to take their own lives—circumstances that would be scrutinized in great detail for the first time by writers in the Japanese colonial period.

Japanese Colonialism: Biopower and the Rise of Moral Statistics

During the Japanese colonial period (1910–1945), a modern medical discourse emerged that would not only challenge traditional explanations and "sanctioned" suicides but also be adopted and internalized by the Korean press and carry unintended consequences. By the 1920s, new nosological and gender labels became prominent in print culture in tandem with the birth of the clinic. Nervous disorders came into the field of vision of public opinion in an era when the fear of infectious and parasitic diseases was becoming increasingly widespread. By affixing psychiatric labels to certain behavior, experts often lumped together symptoms that ordinary Koreans had not understood as a singular, coherent condition, let alone as a medical illness. Biological reductionism, which provided a proper descriptive basis for normative judgment, replaced the complex vernacular vocabularies of afflictions; the public was made to believe that the mind was subject to disease, debility, or even death.

Classified by etiology and pathogenesis, psychiatry offered a medical language to frame and quantify new social pathologies linked to the debilitating "maladies of the nerves." New terms such as the hysteric and neurotic, as well as other types of social pathologies and abnormalities like the "suicidal," embedded in new fields such as psychiatry and criminology, became part of the public discourse. In an era of growing empiricism, the power of the medical gaze provided unprecedented possibilities for a professional group of medical "experts" to intervene and exclusively administer services and sequester populations in institutions in the name of social security. Likewise, national figures on the social conditions—in particular birth and death rates, crime, and suicide—began to be published annually by the Government-General, providing a useful index for gauging the "well-being" of society as "modern technologies of power," which ranged from medical examinations to diagnostic practices began to influence how Koreans thought about suicide, especially as the press adopted this medicalized language and concepts.[8]

A year after conducting its first national census, the Government-General published a detailed fifteen-year report on the demographic trends on the peninsula titled "Population Phenomena of Korea in 1926," which also included a section on suicide rates in the colony.[9] Categorized by race, age, and gender, police suicide reports and tallied annual figures also provided and identified personal motives for committing suicide, which included mental illness, chronic illness, livelihood difficulties, crime of passion or jealousy, shame about crimes committed in the past, discord with relatives, fears about being caught for a criminal act, worries about the future, loss in one's business, difficulties in repaying a debt, disciplinary action by an employer, father or other elder, immoral debauchery, extramarital affair, anxiety over pregnancy, elderly concerns, anxiety over marriage, and despondency about a deformity.[10] Although it is easy to conflate and filter the phenomena of suicide through these statistics, these figures by the Government-General's office can shed light on the rate of recorded suicides—that is, a gradual increase from 474 in 1910 to 1,065 in 1920 to 1,536 in 1925. Investigators also observed a gender discrepancy: female suicides outnumbered male suicides from 1910 to 1919 and incidence of suicide was highest during the spring and early summer, which seemed to corroborate newspaper reports.[11] The report also singled out men in their thirties as the most vulnerable age group.[12]

Not surprisingly, despite these revealing statistics, the report did not offer any concrete recommendations or solutions to help avert suicide in the colony, for that was not Japan's concern. Instead, investigators took a macro-approach to examine these demographic trends to show that suicide was a social phenomenon and that rates were inevitably going to increase as Korea developed into a more modern civilization, especially noting higher incidences of suicides among urban dwellers, which echoed the views of nineteenth-century moral statisticians in Europe such as Jean-Etienne Esquirol and Adolphe Quetelet.[13] For the colonial authorities, these statistics demonstrated that their project of uprooting Koreans from their passive, traditional ways was indeed on the road to success.

The colonial state co-opted the police and medical community to deal with suicide. In the first instance, the police began to analyze the victim's suicide note and interrogation of family members to create diagnostic labels and classifications. As in the Weimar Republic, where "social prejudices and clichés about suicide held by the criminal police and the public certainly played a great role in determining whether a given death was classified as murder or suicide," the police also drew on public opinion to determine whether the victim had had a mental disorder.[14] They treated failed suicide as a criminal act, often subjecting the survivor to interrogation. In cases of unnatural deaths or discovery of an unidentified corpse, a doctor always accompanied a policeman to the scene of the crime. When the doctor provided his expert opinion that the death was a suicide, the police would then contact a family member and arrange for the transfer of the body and permission to proceed with the burial.[15] However, if the police or doctor suspected foul play or considered it a homicide, the authorities conducted an autopsy as part of the investigation process.[16] For example, on October 29, 1929, the *Tonga ilbo* reported that a railroad employee had discovered a corpse floating on the Han River. During the autopsy, the police found a suicide note in the victim's pocket, which identified the man as Hwang In-su, a thirty-eight-year-old who had become despondent after being laid off from work. After interviewing the victim's friend Kyŏng Su-yŏng, the police informed the press that Hwang had experienced a series of setbacks, including the contraction of pulmonary tuberculosis and the loss of his job, which prompted him to hurl himself into the river.[17]

The Government-General required all of its police precincts to tally the number of completed suicides, which also included documenting each victim's age, race, gender, and cause of death. It did not, however, include failed suicides. According to their annual reports, the most common methods were hanging, followed by drowning, consuming poison, and hurling oneself in front of a passing train.[18] Sodium hydroxide, also called lye or caustic soda and a popular whitening detergent introduced to Korea by the Japanese, was the most common poison, followed by rat poison, morphine, bittern, soy sauce, corrosive sublimate, petroleum, blowfish eggs, brine, and a lethal medicine called Nice.[19] Women often chose indirect methods that did not involve physical violence. Their male counterparts opted for more violent methods, reflecting stereotyping in the press.

Suicidal Behavior and Popular Discourse

As publication laws eased in the aftermath of the March First Movement in 1919, suicide featured quite regularly in the popular press, which had multiple agendas for its exploitation of the public's morbid fascination with violence and tragedy with its catchy headlines.[20] It also took advantage of its free access to information; in contrast to privacy laws today, private information was not restricted, so the press could obtain information from the local police stations and publish a biographical sketch of each suspect or victim that included his or her name, gender, age, and address. One role the press sought to play was to modernize its reader's understanding of suicide, eschewing traditional explanations. It also exposed readers to new medico-legal terms such as *yŏmseju'ŭi* (tedium vitae), *ch'ijŏng* (crimes of passion), and *chŏngsin isang* (mentally ill) to explain the crisis of selfhood in the modern age. Fiction writers, who often serialized many of their stories in the newspapers, contributed to this effort, often experimenting with new genres such as realism to define a wide range of "modern" emotions such as amorous passion or unrequited love. This is not to suggest that people had not experienced these emotions earlier; in the newspaper, however, they were linked to problems of modern life.

Moreover, these stories grappled with the price of pursuing individual happiness at the expense of the family. Indeed, the ideals of romantic love and fulfillment were quite different than a decade earlier when the long-standing neo-Confucian ethos of preserving chastity or family honor was the conventional trope in any narrative discourse.[21] If the state endorsed the idea that social forces more significant than the individuals were the catalysts for suicide, writers focused on the everyday life and the changing consciousness of the people, especially themes of self-discovery and the free-willed individual.

The press also focused on specific suicide cases committed in public spaces, which gained particular notoriety. On October 7, 1917, city planners at the Government-General's office celebrated the completion of the Han River footbridge that linked Yongsan to Noryangjin. Despite the great fanfare by the state, the bridge had suddenly become the venue of choice for suicides, tarnishing the glory of the colonial authorities for its engineering achievements.[22] The well-worn saying "I am going to the Han River" also signified something pathological, evoking feelings of sadness and despair, captured so brilliantly by the poet Kim Ki-rim.[23] The reports of daily suicides in the press prompted the Yongsan Young Men's Association to hold a *siagwi,* a special Buddhist rite that offers food to hungry ghosts, to appease the homeless spirits.[24] Unable to curtail the growing number of suicides on the bridge, the Yongsan precinct even sponsored an event offering to reward a prize to the person who could come up with the most catchy suicide prevention slogan. Of the potpourri of entries, ranging from the poetic to the religious, "Wait a minute and restrain yourself" received the most votes. To further promote this campaign, the state, frustrated by the negative publicity that the bridge brought to its reputation, opened a counseling-cum-employment center next to the bridge.[25] Despite this publicity stunt, suicides continued to increase, which prompted the Yongsan precinct to solicit funds from a Japanese expatriate, who offered financial assistance to help defray the costs of installing lamps, railings on both sides, netting, and a full-time watchman. Nonetheless, the number of suicides on the bridge kept increasing every year.[26]

The press sought to understand the new triggers for suicide in modern society; these were linked directly to the process of colonial modernization, which had an impact on economic pursuits. The press cited the stresses of modern life—*pigwan* (feelings of despondency) and *saenghwallan* (livelihood difficulties) as the primary causes of suicide in colonial Korea.[27] Contemporaries warned that rapid industrialization, which had triggered an outmigration of people from the countryside to the urban centers, had already transformed into an acute social crisis as traditional social relations were quickly unraveling. However, as more people faced isolation and anxiety with the breakdown of traditional social structures and class expectations, which often resulted from a lack of normlessness and social regulation, they experienced a sense of anomie (in the Durkheimian sense) that could induce suicidal tendencies.[28]

The press profiled the rise of a new group at risk for suicide—namely, the *lump'en* (unemployed intellectuals who drifted around in the urban centers without a sense of belonging or community). After the March First Movement, Im Ki-sun, a twenty-one-year-old male who lived a life of paranoia after escaping arrest, became increasingly frustrated by his inability to exercise his thoughts freely. Despondent about his life, he penned a suicide note describing his brushes with the law and the reasons for his involvement in the youth movement that undermined his parents' wishes and responsibilities to his studies.[29] In the face of high expectations and desire to gain admission to a top high school or university, the press also reported an increase in the number of suicide cases among students over failing the entrance examination, getting caught cheating, doing poorly in school, or feeling pressure to drop out because of financial problems at home.[30] For example, Min Byŏng-ho, an eighteen-year-old student who had recently been admitted to a higher common school in Seoul, became despondent after his parents told him that they could no longer afford to pay for his tuition; he then attempted to commit suicide by jumping off the Han River Bridge.[31]

The press also blamed failures in entrepreneurial ventures as a relatively new trigger for suicide. Fueled by dreams of striking it rich, waves of people from the countryside teemed to find better employment or start up a business. After making a substantial windfall profit in the speculation of rice, a certain Yun Ki-sŏn suddenly suffered a reversal of fortune and lost his entire twenty-year savings in one risky bid. Overwhelmed by his losses, he decided to commit suicide.[32] In another case, the entrepreneur Yi Hwa-il, who had invested his life savings in the beef and pork business, could no longer handle his operations and became very despondent after giving up his house to his creditors; he decided to commit suicide.[33]

But it was not merely the failed entrepreneurs, intellectuals, or students who were allegedly at risk: the press also argued that modernization had a catastrophic effect on the lower classes because the gap between the rich and poor was widening further.[34] For instance, Kim Yong-gwan, a thirty-year-old employee at the P'yŏngyang Eastern Rice Mill, out of pure desperation stole a sack of rice from the warehouse while the owner was away. The humiliation of being caught and reprimanded by a member of the owner's family led to deep embarrassment and anguish; he used a cross-shaped object to slit his stomach.[35] In another case, Wŏn Pok-dŏk, the twenty-five-year-old wife of Yi Ch'ang-ol, decided to hurl herself in front of a speeding train after being berated for stealing some bean leaves from her neighbor's field.[36] Although desperate straits had driven both Kim and Wŏn to steal, it was their moral "conscience" that compelled them to commit suicide.[37]

A study conducted by the *Tonga ilbo* of Kyŏnggi Province in 1927 identified poverty as the root cause of livelihood difficulties.[38] Individual cases of North P'yŏngan Province, North Ch'ungch'ŏn Province, and the city of Taegu also reported similar findings.[39] Suicide was seen as a symptom of a broader crisis as unscrupulous landlords continued to charge higher rents, often stripping farmers of their tenancy rights. Water shortage and poor yields coupled with anxieties over debt were cited most commonly as the cause for suicides in the countryside.

This crisis of poverty is reflected in the literature as well. In Cho Myŏng-hŭi's short story "Nongch'on saramdŭl" (Rural people), Wŏn-bo, an honest and diligent farmer, gets into a scuffle with the son of Kim Ch'am-bong, a gendarme, over water for his parched fields and ends up serving time in jail. Not only does he lose his wife to Kim Ch'am-bong's son, he also loses his right to tenancy to an evil landlord after another bad drought. In desperation, he tries to steal from the landlord's house, is arrested, and ends up hanging himself in the holding cell at the local precinct.[40]

The *Tonga ilbo* placed the blame for this destructive poverty squarely on the colonial authorities. For instance, it described a slash-and-burn farmer in North Hamgyŏng Province, who could no longer make a livelihood because of the Government-General's new forest ordinance.[41] The press argued that impoverished families ended up resorting to prostitution as a last resort for survival. For example, in Hong Ch'ang-un's case, his inability to provide from his family compelled is wife, Yi Yŏng-nyŏ, to sell her body for food. Unable to forgive himself, Hong became despondent, drank lye, and died.[42] In another tragic instance, a mother in her thirties, despite using every muscle and joint, felt "disgraced" that she could no longer feed her family and decided to commit suicide.[43] Literary representations such as Yŏm Sang-sŏp's short story "Isim" (Disloyalty) became common. In this story, Ch'ung-kyŏng is sold into prostitution by her alcoholic husband. Unable to escape from this grimness, she starves herself and then swallows sleeping pills.[44]

The press reported stories of not only countless individuals and their struggles with livelihood insecurities but also entire families who committed suicide to escape poverty.[45] Because the large urban centers offered no social safety nets, the elderly were also vulnerable and found it very difficult to adjust to dramatic economic and social changes. The elderly allegedly also committed suicide because of harsh treatment by their "unfilial children," loss of self-esteem, inability to deal with chronic illness, despondence over not having children, and resentment over a spouse's or child's squandering the family's fortunes.[46]

Related to livelihood difficulties were terminal or chronic illnesses, which the press cited as another common motive for committing suicide. For the first time, the press sought to link suicidal tendencies to the chronicity (*mansŏng*) of illness. After she had her finger amputated at a hospital in Hwanghae Province because of boils, Ms. Yi, the twenty-four-year-old wife of Kim Hwan-yŏng, suffered yet another setback when her right wrist had to be amputated. Unable to work and pressed with livelihood difficulties, Ms. Yi, who was already several months pregnant, hurled herself into the river with her two-year-old son. In another case, Ms. Cho, the wife of Kim Ch'un-san, became despondent after contracting venereal disease from her husband. Not only had she lost her youth to this disease, she regretted not marrying the man whom she loved. With her beloved child strapped to her back, she took her life.[47] Contemporaries argued that people with *nanch'i* (an incurable illness)—such as tuberculosis, measles, hemorrhoids, leprosy, syphilis, or epilepsy—were a burden to their families, who were driven to the edge by caring for the afflicted.[48]

Although the press generally cast blame on natural causes, the colonial state, or modernity for some suicides, it took issue with the traditional family, which contemporary reformers viewed as the source of many social evils. They primarily targeted early marriage, arranged marriage, male infidelity, abusive mothers-in-law, violent relationships, the buying and selling of wives, profligate husbands squandering family fortunes, and divorce. The press saw these as practices that compelled women in particular to take their own lives.[49] In some cases, suicide was a "weapon of the weak," a form of protest.[50]

Earlier efforts to abolish early marriage during the Kabo Reforms of 1894 and, more recently the Government-General Civil Code (Ordinance number 13) amended in 1922, which fixed the minimum age for marriage at seventeen for boys and fifteen for girls, had no visible impact in the countryside. The frequency of suicides by young girls became the cause célèbre to put an end to the oppressive practice of early child marriage. Most often, husbands themselves resorted to physical violence to punish their young wives. For example, a fifteen-year-old girl from Ŭiju county who became a *minmyŏnŭri* (young daughter-in-law-to-be) to a twenty-four-year-old man when she was eleven years old, could no longer handle the sexual demands of her husband and fled to another village to work as a maid. She was captured by her husband and taken home. Unwilling to live a life of bondage, she took her own life by jumping off a railroad bridge.[51] In other cases, it was the despotic behavior of the mother-in-law who joined her son in abusing her young daughter-in-law that was to blame. For example, Yi Sŏng-nyo, an eighteen-year-old, who had been forced to marry when she was only thirteen, decided to commit suicide rather than endure abuse by her "heartless" husband and overly critical mother-in-law.[52]

Such reports suggested that women had little recourse to address their grievances and had no choice but to resort to extreme measures to escape their unbearable circumstances. Although it is difficult to ascertain how long women waited before they decided to "end" the abuse and unhappy marriages, writers from this period do shed some light on the motives as they dissected the traditional family and the tenuous relationships between couples, siblings, and *kobu* (daughter-in-law and mother-in-law), perhaps the most prone to conflict and commonly recognized as a serious problem affecting family dynamics. In the majority of cases reported, nagging, meddling, and taunts by overzealous mothers-in-law often provoked young daughters-in-law to commit suicide. In the case of Ms. Chu, the nineteen-year-old wife of Kung Chong-o, a farmer in North P'yŏng-an Province, her request to visit her parents during the peak of the harvest season ended with a harsh reprimand by her mother-in-law. After being scolded for her insensitivity and the timing of her request, Ms. Chu became despondent and decided to hang herself from a mulberry tree.[53]

In some cases, wives publicly challenged the authority of the family by questioning the integrity of her husband or mother-in-law in an oral confrontation followed by a public display of anger. For example, Pak Pong-dŏk, the twenty-one-year-old wife of Yi Hong-yŏng, after a heated argument with her mother-in-law and sister-in-law, decided to commit suicide by hurling herself dramatically into a well.[54] In another case, it was the exasperated mother-in-law who drank poison to humiliate her daughter-in-law publicly.[55] In the most tragic cases, the press highlighted mothers who would carry their children on their backs and plunge into rivers or wells.[56]

The press also highlighted marital discord, especially acts of infidelity, as leading to acts of violence or suicide. Kim Kye-saeng, the feature writer for a women's column in the *Tonga ilbo,* discussed the recent suicide of Han Kyŏng-suk, a mother of a two-year-old boy; she was several months pregnant at the time. The columnist suspected that a letter she had received several days earlier had been written by Han, who explained in detail her decision to end her life rather than granting her cheating husband a divorce. Kim emphatically explained to her readers that any woman would get teary-eyed after reading Han's letter and suggested that "a lot of women would want to do the same thing and end their lives." Moreover, she felt that "this deep-rooted masculine culture of drinking and frequenting taverns" gave Korean women no alternative but to suppress their *pun* (anger) and *ŏgul* (feelings of unfairness). However, Kim noted that "women have a pretty good amount of *cha-a* [self-esteem] and can end their suffering if they wanted to."[57] In other words, a woman could "choose" to die if she wanted to and "sever the ties that [bound her] to others."[58]

The growing rate of family dissolution, especially in urban centers, also concerned contemporaries because of its links to suicide. Although many women wanted to end their unhappy marriages and to pursue their happiness, husbands often ignored their pleas and refused to file divorce papers. To be sure, women had little legal recourse against their husbands. For example, Ch'oe Wŏn-sil's attempts to salvage her marriage fell on deaf ears as her husband demanded a divorce and that she return to her parental home. After being rebuffed by her parents, Ch'oe became despondent and decided to hurl herself over a bridge and drown herself in the river.[59]

As a result of changing ideas about sexual freedom and desires, stories of misguided love affairs and suicide pacts frequently appeared in the press, triggering a "Werther effect" in the 1920s and early 1930s as copycat suicides suddenly became a fad of the times.[60] According to Pak Chong-sŏng's analysis of Government-General statistics, a total of 490 people successfully committed love suicide between 1910 and 1942.[61] Arranged marriage was cited in the press as another common motive among young couples who committed *chŏngsa* (amorous love suicide), especially when one of them was being held hostage to the whims of the family. The press also reported stories of unrequited love, particularly among the *kisaeng* (courtesan) and their paramours. In one incident, Ko Nam-yŏn, a twenty-two-year-old *kisaeng,* decided to poison herself after the parents of her lover, Ch'oe Pyŏng-ho, refused to accept her because of her low social status.[62] Writers also granted greater agency to lower-class women in their novels when they confronted prejudices based on class or simply selfish actions of their clients. For instance, in Hyŏn Chin-gŏn's short story "Kŭrip ŭn hŭlgin nun" (Yearning for his scowling eyes), the protagonist becomes enamored with Ch'ae-sŏn, a *kisaeng,* and accumulates major debts from purchasing expensive gifts for her. Driven to a corner, he decides that the two should commit suicide by consuming opium together. However, Ch'ae-sŏn experiences a moment of clarity about her fate, which is not tied to her lover's financial woes and resolves not to swallow the opium. The protagonist, feeling pangs of conscience, attempts to remove the opium from Ch'ae-sŏn's mouth to save her life only to realize that she had no intention of dying with him. Shocked at this unexpected "betrayal," he stares at Ch'ae-sŏn in anger as he dies slowly from the poison.[63] In this twist ending, the suicide highlights the raw narcissism of a privileged protagonist who was willing to take yet another life to justify his actions. Of course, in other instances in literature, the author punished women for their self-indulgence. For example, in both Na To-hyang's "Ch'ulhak" (Expulsion from school) and Yŏm Sang-sŏp's "Cheya" (New Year's eve), the female protagonists are accused of living dissolute lifestyles, treated as moral degenerates by society, and compelled to write confessional letters before they commit suicide.[64] In general, the tone was one of moral reproach against suicides committed for personal vanity's sake without a thought for the family or Korean society as a collective. The press and literary writers began to construct differences between suicide in response to social pressures and documented mental illnesses.

Mental Illness: The Pathologization of Deviant Behavior

The birth of the mental clinic influenced the development of a medicalized discourse of mental illness in the press in Korea during the colonial period, which resulted from the convergence of several developments. First, a new rhetoric of care emerged in Japan that was critical of the domestic confinement law enacted in 1900 (Seishin byōsha kang ho). The Confinement and Protection of the Mentally Ill Act called on families to incarcerate their afflicted members in their homes. In opposition, advocates such as Shūzō Kure (1865–1932), a professor of medicine at Tokyo Imperial University and superintendent of Sugamo Hospital, argued that changing demographics coupled with a new division of labor made it extremely difficult for families to care for the mentally ill and urged the government to fund and establish public mental hospitals.[65]

Whereas Kure's position reflected a general humanitarian concern to relieve the suffering of the insane, the actions of the colonial authorities reflected their more significant concern with controlling and policing the population. Thus a medico-juridical discourse that supported the confinement of specific individuals in the interest of public security took shape. In other words, the psychiatric ward was not valued as much for its therapeutic effects in the colonies as for its service as a police measure, a custodial act of the state. As in Japan, the traditional practice of concealing the mentally ill behind closed doors made it quite difficult for the colonial state to document and incarcerate them. Quite apart from their desire to preserve family honor, Koreans were naturally suspicious about committing afflicted family members to a facility operated by the colonial state.

The emerging field of psychiatry required specialists, who could now invoke the language of mental illness to explain criminality and suicidal risk. Ordinary people were urged to "seek out" a doctor who specialized in "mental illness," admit "unwanted family members" to the hospital and get them medicated, and to refrain from engaging in superstitious practices of the past, which would only trigger unnecessary "commotions in the household."[66] By the 1920s, the press increasingly began to report that urban diseases such as neurasthenia and hysteria, an archetypical feminine functional nervous disorder, and a well-established diagnostic category in Japan, contributed to suicide among the urban population.[67]

Writers described neurasthenia as chronic fatigue but also the incapacity to deal with a modern and fast-changing world characterized best by the flaneur Kubo, who walks aimlessly around Seoul in Pak T'aewŏn's *A Day in the Life of Kubo, the Novelist.*[68] Although the term "neurasthenia" was first coined by the American psychiatrist George Beard in 1869, it was also seen by contemporaries as a disease common among male intellectuals, professionals, and urban workers because of their busy lifestyles. A Korean specialist characterized the neurasthenic in pathological terms, pointing out that "most are despondent because life is difficult for them" and "they are all suffering from *chŏnsin soran* [disturbance of the mind]," which could trigger them to commit suicide.[69] The press began to accept such explanations and framed suicides in a medical framework: for instance, the *Tonga ilbo* reported that on May 12, 1928, Choi Sŏng-mun, a twenty-seven-year-old student enrolled at Rikkyo University in Tokyo, threw himself overboard in the choppy waters near the Genkai-nada in Kyushu. After recovering some items that Choi had left on board, the police contacted his family in Pusan and concluded that he was not only despondent about his paralyzed right leg but had also been suffering from neurasthenia, which compelled him to quit his studies and return to Korea.[70] Not surprisingly, the pharmaceutical companies jumped on the bandwagon, taking out large advertisements in the newspapers, promising readers effective cures to neurasthenia and other "urban" scourges.[71] The pharmaceutical elements made it into novels as well such as one by Pak T'aewŏn in which a nurse offers the protagonist Kubo a concoction of 3B 笈, a placebo of water mixed with three bromides to help him with his headaches.[72]

Between 1914 and 1928, 116 Korean women were admitted to the Government-General Hospital's psychiatric ward to be treated for hysteria.[73] Most were clinically identified as neurotic, psychotic, or frigid and subject to a wide range of specialized treatments specific to her reproductive organs.[74] The press also featured several stories of hysteric women in the urban centers to inform the public about this new female malady associated with the "nerves." For example, a certain Yang Sŭng-hyŏk claimed that he had a loving relationship with his twenty-two-year-old wife, Kim A-chin, and their son until she was diagnosed with hysteria; she then began to express a desire to die frequently. One evening, a family member caught her drinking lye and rushed her to Chŏnchi Hospital for emergency treatment.[75]

Some writers argued that external circumstances such as chronic illness or economic hardship could contribute to a person's development of mental illness. For example, Song Sang-ok, a sixty-seven-year-old man in North P'yŏngan Province, was apparently living a comfortable life until his diagnosis with a chronic stomach disorder and respiratory disease. According to the newspaper report, the intense pain began to affect his nerves, leading to his drowning in a pond near his home.[76] Whether it was by hanging, drowning, drinking poison, or hurling oneself in front of a speeding train, the press identified a common pattern whereby the victim first received a diagnosis of a disease that then triggered mental illness and suicide.[77] However, although it is easy to identify the motive for suicide, no one could accurately predict when the afflicted would experience the final "break," such as that of Kim Ŭng-su, a twenty-five-year-old man who had become despondent. Although his family members were vigilant and made sure he was not left unattended because of his "mental illness," one day Kim "suddenly" slashed his stomach with a knife, spilling his intestines, and died from the wounds.[78]

In other instances, the press attempted to find the roots of mental illness and suicide in domestic issues. In 1922, it reported the case of Yun Ch'i-ik, whose husband had not sent a single letter to her after he went to study in the United States. She allegedly became mentally ill in response this lack of communication and drowned herself in a small pond near her home.[79] In such cases, the press offered a causal explanation, linking some internal family crisis to mental illness, which then induced suicide, shifting the blame from purely political or socioeconomic forces to individuals who could have prevented the tragedies.

The press also expressed sympathy for family members on whom the burden of caring for the mentally ill fell. It often highlighted the plight of spouses who were the most vulnerable to acts of violence or becoming victims themselves. One option to was commit an individual to the hospital, but fears of social stigma—which could potentially jeopardize marriage prospects—compelled many to conceal their ill family members behind closed doors. This practice affected married women more than men given gender expectations that a wife remain at home, as was the case of Kim Bok-sil, who decided to drink lye to escape the torment of living with her mentally ill husband and oppressive in-laws who refused to release her.[80] Other wives met a violent death at the hands of their ill husbands: Yi Sil-dan, a twenty-one-year-old, was stabbed in the chest with a kitchen knife because her husband was angry that she had become cold toward him; before the murder, he had expressed his fears that she was going to marry someone else.[81]

Concerns that their mentally ill relations were a danger to themselves and perhaps to the public is reflected in the press. When an afflicted member escaped from home, families routinely filed the disappearance to the police for fear of violence either to the person's self or the public. In Yangdŏk county in North P'yŏngan Province, Yi Pong-hwa, the wife of Yi Yi-chŏng, reported her husband missing at the local precinct. She told the authorities that he had had some kind of mental illness and had been missing for several days. After searching the county, the police discovered Yi, who had hung himself in a pine forest, and after an autopsy called Pong-hwa to claim his body.[82] In all criminal cases, the police released information to the public after conducting its investigation, which usually took two to four days. If no suicide note was found, the police interviewed family members and neighbors to determine whether the victim had any mental disorders. When the police did not release an autopsy report, the press often speculated and informed its readers that "in all likelihood, the cause was mental illness."[83]

Writers tried to link those suffering from *chŏngsin isang* (mental disorder) with destructive and harmful behavior, mirroring symptoms of those in the asylum, who were deemed at risk of destroying themselves and others. The "mad" would suddenly experience confusion in the mind, which would trigger them to behave in bizarre and unpredictable ways. Suicide reports of patients institutionalized in the infamous Eighth Ward of the Government-General Hospital made headlines, such as Yŏng Sun-nam, a nineteen-year-old from Chinju, who hung herself before the attendant nurse could rescue her, nothing was more frightening than stories of "crazy people" confined in their homes waiting to kill themselves or others.[84] In such cases, the press observed that the families had to be vigilant to make sure the mentally ill would not inflict harm to themselves or other family members.

All these stories helped shape understanding and practices of suicidal behavior among the local population, but they were woefully deficient in providing information on how to deal with the afflicted. However, they do show how the emerging field of psychiatry, which offered a medical language to frame and quantify suicide as pathological, often competed with sociological explanations, which linked it to changing socioeconomic conditions.

Conclusion

As this chapter shows, stories of people committing suicide entered the public discussion for the first time in the 1920s as newspapers started to collate and analyze statistical evidence offered by the Government-General's office. Contemporaries drew on various assumptions as they tried to rationalize the motives for self-destruction. The Korean press treated suicide as a social phenomenon, a tragic by-product of colonialism and a flawed modernity, which transformed existing social mores and economic circumstances. At the same time, it paradoxically used suicide to critique institutions like the traditional family as well as explain the social problems of the day.

On the other hand, the colonial state saw no cause for alarm in the increase of suicide rates and argued that Korea, like Japan and other "civilized nations," had to pay the price to "become modern," and that suicide was one of the destructive aspects of civilization. At the level of governance, these statistics were considered useful indices for gauging the "moral well-being" of Korean society. Using the census, the state could analyze who was committing suicide and their motives. If the press used these statistics and police reports to critique the fast-paced momentum of modernity and its corrupt values, it also advocated for significant medical intervention to prevent family tragedy, which would have to come at a steep price for Koreans that would involve greater interference and surveillance by the colonial state apparatus.

Finally, as new medical truths came to be formulated about mental illness, the emerging field of psychology provided a language to frame particular suicides in pathological terms. Of course, this kind of logic suggests that questions of illness and pathology possessed "an obvious rationale," and such "assumptions embedded within these approaches" and their "effects" were never questioned or analyzed.[85] Despite these limitations, this growing widespread awareness as a socially borne pathology, as well as an eagerness to explain self-destruction within such conceptual frames, shaped the meaning of suicide in colonial Korea.

Notes

Kim Ki-rim, "Han'gang indogyo" [Han River footbridge], in *Kim Ki-rim Chŏnjip* 1 (Seoul: Namsŏldang, 1988), 101.

1 *Tonga ilbo,* June 27, 1925.

2 Pak Chong-sŏng, *Sarang hada chukta: Chŏngsa ŭi chŏngch'ihak* [To love and then to die: The politics of love suicides] (Seoul: In'gan sarang, 2012), 156–157.

3 Pak Chong-sŏng, *Sarang hada chukta,* 83–85; Chŏng Ku-sŏng, *Chosŏn ŭi memento mori: Chosŏn i pŏrin chadŭl ŭi chu'gŭm ŭl kiyŏk hara* [Chosŏn dynasty: Memento Mori: Remembering the deaths of those who were discarded by the Chosŏn dynasty] (Seoul: Aepŭlbuksu, 2010); and Min Sŏng-kil, *Hwabyŏng yŏn'gu* [Study of fire illness] (Seoul: ML, 2009), 16–18.

4 Na Yŏng-in, "Chasal kwa chŏngsin chilhwan, kŭrigo Chosŏn ŭi kŭndaehwa" [Suicide and mental illness, and Korea's modernization], *Sŏul Taehak Sinmun,* March 17, 2012.

5 Ibid.

6 Pak Chong-sŏng, *Sarang hada chukta,* 83–87.

7 The Committee for the Compilation of the History of Korean Women, *Women of Korea: A History from Ancient Times to 1845,* translated by Yun-Chung Kim (Seoul: Ewha Womans University Press, 1976), 104–105.

8 Ian Marsh, *Suicide: Foucault, History and Truth* (Cambridge: Cambridge University Press, 2010), 220.

9 Chōsen Sōtokufu, *Chōsen Jinkō Genshō* [The population phenomena in Korea] (1926), 422–430.

10 Ibid.

11 *Tonga ilbo,* June 27, 1925.

12 Chōsen Sōtokufu, *Chōsen Jinkō Genshō,* 442–430.

13 Howard Kushner, "Suicide, Gender and the Fear of Modernity in Nineteenth-Century Medical and Social Thought," *Journal of Social History* 6, no. 3 (1993): 464.

14 Christian Goeschel, *Suicide in Nazi Germany* (Oxford: Oxford University Press, 2009), 47.

15 *Tonga ilbo,* December 31, 1924.
16 Ibid., August 18, 1927.
17 Ibid., October 29, 1929.
18 Ibid., May 6, 1923.
19 Ibid., November 14, 1921; December 28, 1921; May 13, 1922; June 18, 1923; September 24, 1923; October 26, 1923; October 30, 1923; April 28, 1925; November 12, 1928.
20 Ibid., February 22, 1924; February 25, 1924; August 3, 1924; September 3, 1925.
21 Yi Yŏng-a, "1920-nyŏndae sosŏl ŭi 'chasal' hyŏngsanghwa yangsang yŏn'gu" [A study on the figurative methods and meanings of 'suicide' in the 1920s novels], *Han'guk hyŏndae munhak yŏn'gu* 33 (April 2011), 207–248.
22 *Tonga ilbo,* July 2, 1920.
23 *Chung-ang ilbo,* October 12, 2010.
24 *Tonga ilbo,* July 28, 1921.
25 Ibid., May 16, 1922.
26 Ibid., August 18, 1923.
27 Ibid., August 5, 1926.
28 Emile Durkheim, *The Division of Labor in Society,* trans. W. D. Halls (New York: Free Press, 1984), 201–212.
29 *Tonga ilbo,* June 20, 1923.
30 *Chungwae ilbo,* November 14, 1929; *Sidae ilbo,* May 15, 1924; *Tonga ilbo,* March 8, 1922; February 25, 1924; November 18, 1924; April 13, 1925; July 30, 1926; February 22, 1927; March 17, 1928; March 25, 1928.
31 *Tonga ilbo,* May 15, 1924.
32 Ibid., January 4, 1925.
33 Ibid., July 3, 1924.
34 Ibid., June 4, 1924.
35 Ibid., December 15, 1925.
36 Ibid., August 25, 1925.
37 Ibid.
38 *Tonga ilbo,* April 10, 1923; June 15, 1923; March 1, 1925; June 9, 1927.
39 Ibid., April 13, 1925; July 7, 1927; October 10, 1928; March 4, 1929.
40 *Hyŏndae Pyŏngron* (November 1926).
41 *Tonga ilbo,* June 1, 1925.
42 Ibid., August 23, 1924.
43 Ibid., March 1, 1925.
44 *Maeil Sinbo,* October 22, 1928 and 4 April 24, 1929; *Tonga ilbo,* December 16, 1925; May 14, 1926; June 1, 1922; August 23, 1924; February 24, 1925; June 1, 1925; July 18, 1926; December 14, 1926; July 30, 1927; May 16, 1928; June 6, 1928; May 31, 1929; November 30, 1929.

45 *Tonga ilbo,* December 16, 1925; May 14, 1926.
46 Ibid., June 1, 1922; August 23, 1924; February 24, 1925; June 1, 1925; July 18, 1926; December 14, 1926; July 30, 1927; May 16, 1928; June 6, 1928; May 31, 1929; November 30, 1929.
47 Ibid., September 26, 1925.
48 Ibid., May 21, 1921; August 7, 1921; August 13, 1921; November 4, 1921; July 13, 1922; February 7, 1922; August 31, 1923; June 17, 1925; September 26, 1925; November 30, 1927; February 24, 1928; August 24, 1928; July 13, 1929.
49 Ibid., February 8, 1929.
50 See James Scott, *Weapons of the Weak: Everyday Forms of Peasant Resistance* (New Haven, CT: Yale University Press, 1985).
51 *Tonga ilbo,* July 13, 1934.
52 Ibid., May 28, 1924.
53 Ibid., May 19, 1922.
54 Ibid., November 26, 1926.
55 Ibid., January 29, 1923.
56 Ibid., July 21, 1924.
57 Ibid., June 23, 1926.
58 Lisa Lieberman, *Leaving You: The Cultural Meaning of Suicide* (New York: Ivan R. Dee, 2003), x.
59 *Tonga ilbo,* August 23, 1924.
60 Ibid., August 5, 1926.
61 Pak Chong-sŏng, *Sarang hada chukta,* 156–157.
62 *Tonga ilbo,* December 11, 1925.
63 Hyŏn Chin-gŏn, "Kŭrip ŭn hŭlgin nun" [Yearning for his scowling eyes], *P'yehŏihu*, January 1924, https://nocopyright.tistory.com/105.
64 Yi Yŏng-a, "Nuga dŏ mullanhan'ga?" [Who is more promiscuous?], *Chung'ang ilbo,* December 8, 2011.
65 Yi Pu-yŏng, "Ilcheha chŏngsingwa chillyo wa kŭ pyŏnchŏn: Chosŏn ch'ongdokpu ŭiwŏn ŭi chŏngsingwa chillyo (1913–1928) rŭl chungsim ŭro" [With special reference to the psychiatric care and its change under the Japanese colonial rule: Clinical activities at the government general hospital, 1913–1928], *Ŭisahak* 3, no. 2 (1994): 153–154.
66 *Tonga ilbo,* December 18, 1930; and *Chosŏn ilbo,* January 16, 1923.
67 Junko Kitanaka, *Depression in Japan: Psychiatric Cures for a Society in Distress* (Princeton, NJ: Princeton University Press, 2012), 57–65.
68 Pak T'aewŏn, *Sosŏlga Kubo-ssi ŭi iril* [A day in the life of Kubo the novelist] (Seoul: Muhak kwa chisŏngsa, 2005).
69 *Tonga ilbo,* March 1, 1925.
70 Ibid., April 29, 1928.

71 One pill company peddled its popular medicine (*mach'isu*) promising its effectiveness in curing neurasthenia very easily. See *Tonga ilbo,* May 18, 1923.
72 Gu Tul-lae, "Tanchang chipko kongch'aek tŭn Kubo ssi kŏt'gi," *Hankyŏre 21* (October 8, 2010), http://h21.hani.co.kr/arti/culture/culture_general/28237.html.
73 Yi Pu-yŏng, "Ilcheha chŏngsinkwa chillyo," 163.
74 Phyllis Chesler, *Women and Madness,* 3rd ed. (New York: Four Walls Eight Windows, 1997), 93.
75 *Tonga ilbo,* October 5, 1926.
76 Ibid., July 25, 1926.
77 *Chungwae ilbo,* May 23, 1927; *Tonga ilbo,* November 25, 1926.
78 *Chungwae ilbo,* May 23, 1927; *Tonga ilbo,* November 25, 1926.
79 *Tonga ilbo,* August 3, 1922.
80 Ibid., November 21, 1926.
81 Ibid., October 15, 1929.
82 Ibid., September 23, 1926.
83 Ibid., November 23, 1926.
84 Ibid., January 9, 1925.
85 Marsh, *Suicide,* 4.

CHAPTER 5

In Search of an Anticommunist Nation

The World Health Organization and Public Health Planning in Postwar Korea

Jane S. H. Kim

On May 2, 1953, at the Sixth World Health Assembly held in Geneva, Switzerland, the World Health Organization (WHO) revealed its public health rehabilitation plan for the Republic of Korea.[1] Titled "The Report of the WHO/UNKRA Health Planning Mission in Korea," the report was based on the survey of health conditions that the WHO carried out at the request of the United Nations Korea Relief Agency (UNKRA) for two months in South Korea during the summer of 1952. Headed by George MacDonald, the first chair of tropical medicine at the London School of Hygiene and Tropical Medicine and a member of the WHO expert committees on malaria and environmental sanitation, the Korean mission was dispatched on July 3, just a week after Joseph Needham, the renowned biochemist and Sinologist, began his infamous bacterial warfare investigation in China and North Korea.[2] After the initial briefing at the WHO headquarters in Geneva with Brock Chisholm, the former deputy minister of health and welfare of Canada and the first director-general of the WHO, the Korea mission team was given further instructions at the UNKRA headquarter in New York and then at the Western Pacific Regional Office (WPRO) of the WHO in Manila, and arrived in Seoul on August 8. The team stayed in the country until October 8, and the final report was submitted to the WHO on November 24 of the same year.[3]

Considered the blueprint for public health rehabilitation and reconstruction of postwar Korea, the 1952 WHO/UNKRA report marked the beginning of the involvement of the United Nations (UN) and, ultimately, the remaking of South Korean public health. First intended as emergency measures necessary to secure a minimum standard of public health that would prevent further trouble in a Cold War hot zone, the report argues that efforts, such as the one initiated by the WHO in 1952, helped transform the Republic of Korea into a bulwark of anticommunism. With technological expertise and material assistance, the various UN organizations—Food and Agriculture Organization (FAO), WHO, United Nations Educational, Scientific and Cultural Organization (UNESCO), and United Nations International Children's Emergency Fund (UNICEF)—and its donor nations, namely, the United States, sought to transform the Korean peninsula into the very first defense line against communist encroachment in Northeast Asia. To reconstruct the former Japanese colony into a strong anticommunist nation, the WHO Mission drafted a plan for postwar public health, which the United Nations hoped would enable Koreans to become healthy and able-bodied citizens of the Free World. By examining the 1952 WHO public health planning mission to Korea, this chapter illustrates the close connection between the geopolitics of the Cold War and the impact it had on the making of public health in postwar Korea. In accounting the historical context of the 1952 WHO mission, the chapter demonstrates the role of geopolitics in determining the priorities of public health in the postcolonial South Korean nation-state.

Precursor to the Mission: The Bacterial War in the Peninsula

The 1952 WHO mission to Korea came at a time when accusations and denials over the possible use of deadly germs against North Korean and Chinese civilians by the United States were flying back and forth across the Pacific. Called bacterial or germ warfare (*segyunjŏn,* 細菌戦) in the North Korean and Chinese literature, the incident created paranoia over the impending outbreak of massive fatal epidemics in the region. Although the United States, the United Nations, and the WHO repeatedly denied the accusations launched by the two new Asian socialist states, throughout 1952, as the WHO mission unfolded in South Korea, the Chinese and North Koreans mobilized friendly international opinions to challenge the UN and the US intervention into the continuing war on the Korean peninsula. To these two nations, the asserted US use of "germ bombs" or biological weapons against the North Korean and Chinese populations was irrefutable proof of the criminality of US and UN aid to South Korea and intervention into a civil war. To further incriminate the United States and the UN, the two nations even went so far as to ask Joseph Needham, a renowned Sinologist and biochemist who first introduced Chinese science to the West, to organize an international committee to investigate and expose the heinous war crime committed by the United States to the world.[4]

Needham, who was privately skeptical, nonetheless agreed to head the commission, formally titled the International Scientific Commission for the Investigation of the Facts Concerning Bacterial Warfare in Korea and China (ISC) and the timing of their overture in China and North Korea coincided with the WHO Mission team's trip to Korea.[5] On an almost back-to-back schedule, the two sides across the 38th parallel made fact-finding trips over health conditions in Korea. The accusations regarding the use of biological warfare was never verified.[6] However, the heightened hysteria and paranoia over the possible impending health crisis compelled the WHO, the United Nations, and ultimately the United States to see rehabilitation and reconstruction of public health in South Korea not simply as a matter of health concern alone, but also of how health issues were deeply tied to the geopolitics of the region.[7] This recognition of the role that health plays in the rehabilitation and reconstruction of Korea as an anticommunist nation and a member of the Free World could be seen in the programs of public health planning and reconstruction that the WHO drafted after finishing their fact-finding mission.

Public Health Planning for Postcolonial Korea

The final report of the 1952 WHO Mission to Korea was submitted to the director-general of the WHO on November 24, 1952. Running 104 pages, the report was written by the three members of the WHO mission team who had varied backgrounds in medical and public health research. The team consisted of George MacDonald, malariologist and a member of the WHO malaria and environmental sanitation committees, W. G. Wickremesinghe, director of Department of Public Health of Sri Lanka and expert consultant for the Southeast Asian Regional Office of the WHO, and William P. Forrest, a member of the WHO board of finance. For two months, from August to October of 1952, the three traveled throughout the southern part of the country, visiting all major medical institutions and public health facilities in the central, provincial, and local areas. The mission also held a series of meetings with South Korean public health administrators and consultations with foreign health experts working for the UN and other voluntary agencies in Korea. After their survey, they divided the public health planning of Korea into two sections—immediate- and long-term health goals. The immediate goals addressed the pressing problems of epidemic outbreaks and health issues related to the war, and the long-term goals consisted of plans to reestablish the damaged public health and medical system.[8]

For the immediate health problems facing the country, the authors found that since the exit of the Japanese colonial health experts in 1945, the ensuing collapse of the economy and the extant public health system had turned Korea into a breeding ground for mass epidemics. The war worsened the already critical situation by escalating deaths and large movements of people, which made it easy for infectious diseases to rapidly spread into the greater population. Compounding the problems, the destruction of health and sanitation facilities created acute shortages of medical services and necessary drugs. As result, on investigation, the authors found a chain of epidemics—typhoid fever, dysentery, typhus, smallpox, diphtheria, and encephalitis, to name a few—to have broken out in the country. In addition, "large sections of people" suffering from gross malnutrition, rise in mortality due to a rapid spread of tuberculosis, and deterioration of the overall health of the population were found to be sources of pressing concern as well. The authors commended the United Nations Civil Assistance Corps in Korea (UNCACK) for the efforts made to prevent the spread of epidemics in the early days of the war. Yet, while praising the UNCACK for its heroic attempts to tackle complex public health problems, they nonetheless found the UNCACK measures to be "on a scale sufficient to maintain life but [an] inadequate standard for good health." The war now mostly confined along the 38th parallel, the writers hoped that a more permanent and long-term plan for health could be instituted.[9]

According to the WHO mission team, long-term public health rehabilitation was twofold: rebuilding public health infrastructure and establishing systematic control over selected diseases. For public health infrastructure, the report recommended building public health dispensaries and increasing environmental sanitation facilities such as water works and latrines. The health dispensaries were intended to alleviate the heavy concentration of patients in the already crowded hospitals and to provide health service to "no doctor areas," regions of the country where access to health service was hard to come by. Given MacDonald's background in environmental sanitation, the report also paid particular attention to the state of environmental sanitation, pointing to the lack of sanitation facilities, which made increased parasitic infection across the population possible.[10] For the second part of the plan, establishing systematic control over selected diseases, the mission team targeted four selected diseases as " of major social importance"—leprosy, parasitism, tuberculosis, and venereal diseases. Unlike cholera and typhoid fever, which caused infection within a relatively short time, these four diseases had comparatively lower communicability and were not as contagious nor as fatal as plague or hemorrhagic smallpox, which caused thousands of deaths during the Korean War.[11] In fact, the selection of the four diseases showed that the function of long-term goals of public health in Korea was to promote social stabilization that would eventually beget the political stability of the nation.

The use of public health as an instrument for materializing social stability is clear in the selection of leprosy, which is an infectious disease caused by the bacilli *Mycobacterium leprae* that affects the peripheral nerves, mucosa of the upper respiratory tract, skin, and eyes. Its symptoms typically include loss of sensation, skin lesions, dry scalp, loss of hair, loss of eyesight, in some cases loss of retina, reddish skin or thickening of skin and peripheral nerves, and disfigurement of facial bones, usually the flattening of nose due to collapse of nasal cartilage. For some extreme cases, it may result in the loss of limbs due to repeated injuries or infections that might have gone unnoticed or untreated for a prolonged period. Leprosy is known to be transmitted through skin on close contact as well as through the respiratory system. Although it is generally understood that leprosy is transmitted through contact with the infected in a closed environment, the disease has not shown to be sexually transmitted. Also, the prevailing opinion on transmission route has been transmission through the skin, but recent research shows higher cases of transmission through the respiratory system, usually through nasal droplets. The disease, however, is not highly contagious and carries a low communicability that the immunity of most people is strong enough to eliminate during the early stages of infection before developing into full-blown cases. Some are susceptible to leprosy due to genetic mutation and immune weaknesses of the body.[12]

Pharmaceutical innovations have been made in the treatment of the disease since 1941, when Guy Henry Faget, the director of Carville Leprosarium in Louisiana (US Marine Hospital), discovered the efficacy of promin, a sulfa drug in the treatment of disease.[13] Since Faget's discovery, various sulfa drugs—diasone, dapsone (DDS), clofazimine, and rifampicine—have been manufactured to tackle the treatment of leprosy. In 1981, the WHO ratified the use of multiple drug therapy (MDT), a therapy using mixture of the four drugs mentioned, to tackle paucibacillary (tuberculoid) and multibacillary (lepromatous) leprosy. The WHO recommends six months of MDT, using daily intake of dapsone and monthly intake of rifampicine for treating paucibacillary leprosy, and twelve months of MDT, using daily intake of dapsone and clofazimine, and monthly intake of rifampicine to treat multibacillary leprosy. However, with early detection, leprous infection can disappear in about two weeks after beginning treatment. WHO has freely distributed MDT worldwide since 1997, and MDT has been proven quite successful: relapse of leprosy has found to be rare, and no incidents of resistance to the drugs have been reported. Leprosy is considered one of the more "successful" cases of elimination by the WHO. Further, and contrary to popular perception, medical research in the past six decades has consistently proved that leprosy is a low communicable disease, requiring no draconian measures such as isolation or sterilization.[14]

In Korea, the modern treatment of leprosy in Korea began with the founding of leprosaria in Pusan (1907), Kwangju (1908), and Taegu (1913) by American Presbyterian missionaries. The Japanese colonial government followed in 1916 with the establishment of the Sorok Leprosarium, the first national government facility in Korea. In the treatment and management of leprosy patients, both Christian and Japanese colonial leprosaria practiced lifetime segregation, separation of children born from infected parents, sterilization (vasectomy for male patients and abortion for female patients), and the use of chaulmoogra oil as a therapeutic agent, either orally or by injection. Apart from chaulmoogra oil, measures such as lifetime segregation, separation of uninfected children, and sterilization were gradually implemented over time in Korea. The ideas for these practices have been around worldwide since the early 1900s, but it was not until the eve of World War II that segregation and sterilization became officially sanctioned health practices for leprosy in Korea.[15]

When the WHO mission team visited Korea in 1952, they found that the old colonial treatment and management of leprosy continued. Yet, although critical of these "unscientific" practices, the mission team nonetheless supported the South Korean state's leprosy control activities, including the separation of the "uninfected" children from their "infected" parents.[16] In this, the WHO mission team respected the desires of the South Korean government to use public health policies to secure social stability and state control over society. Indeed, use of segregation as an instrument for social stabilization was illustrated in August of 1945. After enduring years of hardship, including forced labor, physical abuse, and starvation under the Japan during World War II, almost two thousand patients escaped from the Sorok Leprosarium, taking advantage of the administrative vacuum created by the retreat of the Japanese police and military. Soon after, the newly revived South Korean newspapers began to warn the public that these dangerous lepers were roaming freely on the streets and to urge the public to report sightings to the police. Well aware of the potential pandemonium caused by these carriers of fatal disease, the American military government and the succeeding Korean government religiously caught and sent the patients back to the Sorok.[17]

By 1952, world health opinions on leprosy had decisively moved away from the old views. The medical world had come to a consensus that leprosy was not fatal and that, thanks to the development of DDS, it is curable.[18] As result, leprosy patients no longer needed to be segregated behind the walls of a leprosaria. Indeed, questions on the validity of lifetime segregation or complete physical and social isolation of leprosy patients have been raised since the days of the League of Nations Health Organization, the precursor to the WHO (1921–1946).[19] Also, the introduction of sulfa drugs—promin, diasone, and dapsone—was resulting in the rapid decline of the leprosy population worldwide, leading to more questions on the usefulness of segregation policies.[20] In Korea, the new drugs for treating leprosy—diasone and promin—had already been introduced by the returning missionary Robert Manton Wilson, former director of Aeyangwon Leprosarium in Yŏsu, South Chŏlla Province, in 1946. In 1941, he was expelled along with other American missionaries of the Southern Presbyterian Church for their opposition to Shinto shrine worship, and when he was invited back as the leprosy consultant for the American military government, he brought with him the latest drugs available. South Korean newspapers did report on the use of new miracle drugs that could cure leprosy as well.[21] In fact, the 1952 WHO team noted in its report the availability of sulfa drugs in Korea at the time. Yet despite the worldwide changes in medical opinion and the introduction of new drugs, the South Korean state continued to pursue complete isolation of patients and their families from society. The WHO mission team seconded the approach. Acknowledging the infeasibility of keeping patients in government institutions for "all lifetime . . . for both scientific and financial reasons," the report nonetheless made it clear that infected patients and those with severe physical disfigurement that made social rehabilitation impossible should be kept in "colonies." Although the report did not specify a precise definition of "colony," the WHO recommendations on leprosy works in Korea qualitatively differed little from the old segregationist view of physical and social isolation.[22]

Encouraged by the WHO endorsement, for the next three decades and more, the South Korean health authorities faithfully followed the recommendations as outlined in the WHO report. One was maintaining colonies—physical and social isolation of leprosy patients declared clear of the disease. The government named these areas "resettlement villages." This enthusiastic continuation of colonial practices could not have been done without full support of and endorsement by the WHO.

What the 1952 WHO mission team's endorsement of discarded colonial leprosy control policies illustrated was the contradiction and hypocrisy inherent within the WHO and of UN views on "health" for "underdeveloped" and "developed" member nations.[23] That is, for recently decolonized nations and those in geopolitically sensitive areas, hence "underdeveloped" countries such as South Korea, health ultimately had to serve the geopolitical and social stability of the nation. Pursuit of health for the sake of health was not an affordable option. For "developed" nations such as the United States, one of the main sponsoring nations for the UN rehabilitation and reconstruction in South Korea, leprosy patients were no longer required to be physically isolated or quarantined for any given time. Kalaupapa Settlement in Molokai, Hawaii (also known simply as Molokai), is a good example for comparison with Korea. Once known as the world's largest leprosarium and made historically famous by Father Damien, a Belgian Catholic priest who devoted his life to the patients in Molokai, only to succumb to the very same disease as the patients, Molokai was one of the first leprosaria in the world to pursue lifetime segregation—that is, permanent quarantine.[24]

Like Sorok Island in Korea, where the Japanese colonial government pursued an aggressive isolationist practice, Molokai's geographic isolation helped guarantee maintenance of rigorous physical isolation and lifetime segregation for almost a century. Yet, with the development and distribution of DDS, the revolutionary sulfa drug mentioned earlier, the Hawaiian Board of Health began to relax the segregation and, by the end of World War II, began to establish what it called rehabilitation centers for patients who were discharged or had left the Kalaupapa Settlement. By the early 1950s, around the same time as the WHO mission to Korea, the records of the Kalaupapa Settlement show that a significant number of patients began to leave the island and were either housed at the rehabilitation center located in the Pearl City (outskirts of Honolulu) or successfully transitioning into city life in Honolulu.[25]

This stark contrast between the WHO leprosy policies recommended for South Korea and Hawaii's practices show that the discredited old medical practices, even if proven unscientific, as the mission team had admitted in the report, nevertheless needed to be continued because the WHO and by extension UN rationale for health in underdeveloped nations was the geopolitical and social stability of the nation-state and not primarily in the best interests of the health of the population.

This is not to say that the 1952 WHO public health plans were carried out smoothly and encountered no opposition along the way. In 1966, Dharmendra, a renowned leprologist from India, visited South Korea as a WHO short-term expert consultant. His task was to inspect the current conditions of leprosaria and ongoing leprosy treatment in Korea. After his tour, he left behind a blistering criticism of the resettlement villages, the very practice once praised by the 1952 WHO mission team as an economic method for managing discharged patients. Astutely pointing out that these "villages" were nothing more than variation on the old colonies and segregation practices, Dharmendra pleaded for the end to the practice.[26]

Yet by this time, the South Korean state, which had forged a collusive and symbiotic tie with the UN bodies that oversaw its development (FAO, WHO, UNESCO, and UNICEF) for more than a decade, could afford to ignore a lone dissenting voice such as Dharmendra's. Moreover, knowing full well that WHO expert opinion held no juridical authority and because Dharmendra belonged to the WHO's Southeast Asian Regional Office rather than to the WPRO (to which South Korea belonged), his controversial report had very little bearing on their continued practices. To the South Korean government and its leprologists, had Dharmendra been dispatched as the expert consultant from the WHO's Pan American Health Organization (PAHO), of which the United States was a member nation, his stinging words might have had an impact. However, a leprosy expert consultant from India, a fellow decolonized nation struggling like South Korea with underdevelopment and poverty, could have very little to offer in terms of expert advice.

For the South Korean government and leprologists who blithely ignored Dharmendra's 1966 report, their views on health remained in tandem with those on the role and the function of public health as articulated by the 1952 WHO mission team. When Dharmendra pointed to the disparity between the new practice and the old, the South Korean government and leprologists responded by citing the economic and financial strain on the national health budget caused by leprosy control and possible social chaos that could arise by the reintegration of discharged leprosy patients back into normal civilian life.[27] Their fierce opposition shows the degree to which the South Korean government and its public health authorities believed that the goal of public health was to serve the nation-state. Thinking of health beyond the scope of geopolitical and social interests or for the sake of health alone could not be tolerated.

To the South Korean leprologists and other medical experts entrusted with the task of rehabilitating and reconstructing the crippled public health system in postwar Korea, status as an underdeveloped nation surviving by financial and material aid from developed donor nations such as the United States made it all the more clear that the role and function of health was to facilitate the economic development of the nation. Although unspoken, the consensus between the South Korean government and UN bodies was the goal of South Korea's advancement. Becoming a developed nation was a raison d'étre for the Republic of Korea. As a divided nation coming out of a tragic war of ideological contestation, South Korea and its ally the United States had to prove the superiority of anticommunism and the values of Free World over communism and socialist utopia being built on the other side of the demilitarized zone. To that end, the UN bodies and the South Korean government had to undertake economic development as a priority and for economic development in the southern part of the peninsula, the overseers of development needed healthy and literate working bodies. It was for this reason that the WHO mission team articulated its vision for the rehabilitating and reconstructing public health in Korea. The South Korean public health administrators and medical experts, the recipients and beneficiaries of this blueprint for an economically developed nation filled with healthy bodies, could only enthusiastically embrace and religiously follow this vision of public health serving the geopolitical interests of the nation-state for the next decade and more.

Conclusion

In 1960, on the eve of the historic April 19 student uprising, Han Sang-t'ae, the Republic of Korea Ministry of Health and Welfare counterpart to the WHO, wrote a short article in *Pogôn segye* (Public health world), to commemorate the eighth anniversary of the 1952 WHO Public Health Planning in Korea.[28] Han, who would become WPRO regional director in 1989, was a graduate of Seoul National University (SNU) College of Medicine, and received a master's and a doctoral degree in public health at the University of Minnesota. He was fluent in English and during the Korean War worked as translator for UN military medical officers. As English translator, he traveled throughout the country and at one point even participated in UN public health work in Hamhûng, South Hamgyông Province, when UN forces briefly occupied the northern regions. After the UN retreat to the south, Han was stationed at the prisoner of war camp in Kôje Island and there found opportunities to observe and participate in mass vaccination, DDT spraying, and provision of medical relief supplies to refugees coming from the north. After leaving the island, he became involved in the restoration of water sanitation works in Seoul. The restoration of water sanitation was one of the long-term projects proposed in the 1952 WHO/UNKRA public health planning report.[29]

Han joined the Ministry of Health and Welfare after graduation and climbed rapidly within the ministry, taking key posts in public health administration along the way. His fifteen years' work at the ministry touched on all aspects of public health administration—statistics, facilities, planning, prevention, and quarantine, as well as leprosy mobile clinic operations. He served as the director of the Bureau of Public Health before joining the WPRO-WHO in 1970. At the WPRO, he worked in various projects concerning rural health care, infant-maternal health, and family planning, before becoming the regional director of WPRO in 1989. His tenure as regional director was marked by his aggressive push for "eradication" of leprosy, tuberculosis, and acute communicable diseases. He made a name for his efforts in the development and distribution of poliomyelitis vaccine and in 1997, the WPRO became the first WHO regional office (of the six mentioned earlier) to declare eradication of poliomyelitis. Han's polio vaccination campaign laid the foundation for the WHO's Global Programme for Vaccine and Immunization, a world vaccination development and distribution program headed by Lee Jong-wook, a fellow graduate of SNU and Han's junior colleague at WPRO, who would become the director general of the WHO in 2003. Much of the global vaccination work that Lee became known for owed its genesis to the targeted disease eradication and vaccination distribution campaigns first initiated by Han Sang-t'ae at the WPRO.[30]

Han's having witnessed the works of the WHO since the early days of the Korean War meant that his 1960 evaluation of the 1952 WHO/UNKRA Public Health Planning in Korea could only be a positive and a glowing one. In the article, admitting the slow progress made in the rebuilding of public health infrastructure in Korea, particularly in the establishment of public health dispensaries, Han made a point of praising the rapid progress made in the reduction of cases of leprosy, tuberculosis, and parasitic worms, the three diseases identified in the 1952 report as "diseases of major social importance."[31] Han's positive assessment and the contrasting silence on the slow progress in the building of public health infrastructure spoke volumes on the direction the South Korean state took on the role of public health in Korea. To the state, public health was to be defined as targeted campaigns carried out on selected diseases and issues such as family planning rather than in establishing a more permanent and systemic infrastructure for health services. The reason for this outlook lay in the immediate results reaped from selected activities such as disease eradication, which produced visible evidence of the workings of the state. Focused treatment of diseases such as leprosy, tuberculosis, parasitic worm infection, and venereal diseases led to demonstrative actions—physical segregation, mass vaccination, worm elimination, and regulation of bodily behaviors in the eyes of the untrained public.

At the WPRO, Han replicated the same type of targeted campaigns of disease control he first learned at the South Korean Ministry of Health. Indeed, the highlight of his tenure as WPRO regional director was the success of poliomyelitis eradication in the Western Pacific region. The dramatic reduction in the number of poliomyelitis cases was proof of the effective workings of the WPRO in improving the overall health of the region. For veteran public health administrators like Han, the origin of visualizing the efficacy of public health could be traced to the 1952 WHO/UNKRA Public Health Planning Mission to Korea. From his exposure to the WHO's vision and the ensuing UN activities for rehabilitating the war-torn nation, Han came to see the valuable function of public health as an instrument for creating geopolitical stability. Also, having observed firsthand the success and failure of the WHO blueprint for rehabilitation of Korea, Han carried over the more "successful" results gleaned from the original 1952 plan to Manila, site of the WPRO headquarters, and reshaped the role of WPRO as an international health organization focused on disease eradication and selected health campaigns. Originally intended as measures necessary to stop the escalation of Korean War, ultimately the 1952 WHO public health planning helped not only redefine the role of public health as an instrument of the state but also reshape the vision of the World Health Organization in adjudicating health in the postcolonial world order.

Notes

1 World Health Organization (WHO), *Official Record of the World Health Organization No.48: Sixth World Health Assembly, Geneva, 5 to 22 May, 1953—Resolutions and Decisions Plenary Meetings Verbatim Records,* http://apps.who.int/iris/bitstream/1065/85647/1/Official_record48_eng.pdf; and *Report of the WHO/UNKRA Health Planning Mission in Korea,* RG 59 150/71/13/03-04 Box 9, National Archives, College Park, MD.

2 Tom Buchanan, "The Courage of Galileo: Joseph Needham and the 'Germ Warfare' Allegations in the Korean War," *History* 86, no. 24 (2001): 503–522; Ruth Rogaski, "Nature, Annihilation, and Modernity: China's Korean War Germ Warfare Experience Revisited," *Journal of Asian Studies* 61, no. 2 (2002): 381–415; and Simon Winchester, *The Man Who Loved China: The Fantastic Story of the Eccentric Scientist Who Unlocked the Mysteries of the Middle Kingdom* (New York: Harper Collins, 2008), 203–208.

3 "Letter to Dr. Brock Chisholm," *Report of the WHO/UNKRA Health Planning Mission.*

4 Buchanan, "The Courage of Galileo"; Rogaski, "Nature, Annihilation, and Modernity"; and Winchester, *The Man Who Loved China.*

5 International Scientific Commission for the Investigation of the Facts Concerning Bacterial Warfare in Korea and China, *Report of the International Scientific Commission for the Investigation of the Facts Concerning Bacterial Warfare in Korea and China* (Peking, 1952), 4; Commission of International Association of Democratic Lawyers, "Report on U.S. Crimes in Korea, March 31, 1952" (P'yŏngyang: International Association of Democratic Lawyers, 1952); Gavan McCormack, "Korea: Wilfred Burchett's Thirty Year's War," in *Burchett: Reporting the Other Side of the World, 1939–1983,* ed. Ben Kiernan (London: Quartet Books, 1986), 20–203; Günther Wernicke, "The Communist-Led World Peace Council and the Western Peace Movements: The Fetters of Bipolarity and Some Attempts to Break Them in the Fifties and Early Sixties," *Peace &*

Change 23, no. 3 (1998): 265–311; Milorad Popov, "The World Council of Peace," in *World Communism: A Handbook, 1918–1965,* ed. Witold S. Sworakowski (Stanford, CA: Hoover Institution Press, 1973), 488; Lawrence S. Wittner, *The Struggle Against the Bomb: One World or None: A History of the World Nuclear Disarmament Movement through 1953* (Stanford, CA: Stanford University Press, 1993); and US Department of State, "The World Peace Council's 'Peace Assemblies,'" *Foreign Affairs Note* (May 1983): 1–6.

6 Milton Leitenberg, "New Russian Evidence on the Korean War Biological Warfare Allegations: Background and Analysis," *Cold War International History Project Bulletin* 11 (1998): 185–197; Kathryn Weathersby, "Deceiving the Deceivers: Moscow, Beijing, Pyongyang, and the Allegations of Bacteriological Weapons Use in Korea," *Cold War International History Project Bulletin* 11 (1998): 176–185; Ruth Rogaski, *Hygienic Modernity: Meanings of Health and Disease in Treaty-Port China* (Berkeley: University of California Press, 2004), 285–299; Ok Ch'ŏl–ung, "Miguk ŭi ch'ittôli nŭn saenghwahak chônjaeng ch'aektong" [America's enraging scheme to incite biochemical war], *T'ongil sinbo* [Unification Bulletin], June 8, 2015, http://www.uriminzokkiri.com/uri_foreign/tongilsinbo/index.php?ptype=sinbogisa&no=95513; and "Suja tŭl ŭn kobal handa, Mije chimnyakkun ŭi pan illyujŏgin taeryang saryuk choeak ŭl (2)—chinan Chosŏn chŏnjaeng sigi Mije ka kamhaeng han taeryang saryuk manhaeng charyo" [Numbers reveal massacres and crimes against humanity committed by American invading forces (2)—evidence of mass killings carried out by American imperialists during the Korean War], *Uri minjok kkiri* [Among Our Nation], June 25, 2015, http://www.uriminzokkiri.com/index.php?ptype=ugisa1&no=112870.

7 John Farley, *Brock Chisholm, the World Health Organization, and the Cold War* (Vancouver: University of British Columbia Press, 2008), 82–83; and Jeanne Guillemin, *Biological Weapons: From the Invention of State-sponsored Programs to Contemporary Bioterrorism* (New York: Columbia University Press, 2005), 99–105.

8 "Letter to Dr. Brock Chisholm" and introduction, *Report of the WHO/UNKRA Health Planning Mission.*

9 Ibid., 3–4.

10 Ibid., 8–10.

11 Ibid., 55–66.

12 WHO, "Leprosy: Fact sheet, updated April, 2016," http://www.who.int/mediacentre/factsheets/fs101/en/; Koichi Suzuki et al., "Current Status of Leprosy: Epidemiology, Basic Science, and Clinical Perspectives," *Journal of Dermatology* 39, no. 2 (2011): 121–129; American Leprosy Mission, "Leprosy Frequently Asked Questions," http://www.leprosy.org

/leprosy-faqs; David N. McMurray, "Mycobacteria and Nocardia," in *Medical Microbiology,* 4th ed., ed. Samuel Baron (Galveston: University of Texas Medical Branch, Department of Microbiology, 1997); Center for Disease Control and Prevention, "Hansen's Disease (Leprosy)," http://www.cdc.gov/leprosy/index.html; Pieter A.M. Schreuder, Salvatore Noto, and Jan Hendrik Richardus, "Epidemiologic Trends of Leprosy for the 21st Century," *Clinics in Dermatology* 33, no. 1 (2016): 24–31; R.J.W. Rees and A. C. McDougall, "Airborne Infection with *Mycobacterium Leprae* in Mice," *Journal of Medical Microbiology* 10 (1977): 63–68; and Diana N.J. Lockwood, "Human Immunodeficiency Virus and Leprosy: An Update," *Dermatologic Clinics* 29, no. 1 (2011): 125–128.

13 Guy Henry Faget, F. A. Johansen, and Hilary Ross, "Sulfanilamide in the Treatment of Leprosy," *Public Health Report* 57, no. 50 (1942): 1892–1899; and Guy Henry Faget et al., "The Promin Treatment of Leprosy—A Progress Report," *Public Health Report* 58, no. 48 (1943): 1729–1741.

14 WHO, "Leprosy Elimination: WHO Multidrug Therapy (MDT)," http://www.who.int/lep/mdt/en/; R.J.W. Rees, J.M.H. Pearson, and M.F.R. Waters, "Experimental and Clinical Studies on Rifampicine in Treatment of Leprosy," *British Medical Journal* 1, no. 5688 (1970): 89–92; S. H. Yawalkar et al. "Once-monthly Rifampicine plus Daily Dapsone in Initial Treatment of Lepromatous Leprosy," *The Lancet* 319, no. 8283 (1982): 1199–1202; Corinee S.C. Merle, Sergio S. Cunha, and Laura C. Rodrigues, "BCG Vaccination and Leprosy Protection: Review of Current Evidence and Status of BCG in Leprosy Control," *Expert Review of Vaccines* 9, no. 2 (2010): 209–222; Maninder Singh Setia et al., "The Role of BCG in Prevention of Leprosy: A Meta Analysis," *The Lancet: Infectious Diseases* 6, no. 3 (2006): 162–170; WHO, *WHO Study Group on Epidemiology of Leprosy in Relation to Control* (Geneva: World Health Organization, 1985); WHO, "Global Leprosy Update, 2013: Reducing Disease Burden," *Weekly Epidemiological Record* 36, no. 89 (September 5, 2014): 389–400; and Laura C. Rodrigues and Diana N.J. Lockwood, "Leprosy Now: Epidemiology, Progress, Challenges, and Research Gaps," *The Lancet: Infectious Diseases* 11, no. 6 (2011): 464–470.

15 Jane S.H. Kim, "Leprosy in Korea: A Global History" (PhD diss., University of California, Los Angeles, 2012), 9–64, 114–179.

16 "Leprosy," *Report of the WHO/UNKRA Health Planning Mission*, 55–59.

17 Jane Kim, "Leprosy in Korea," 182–204; "Nabyông hwanja rûl ilso, Sorokto ro suyong kyehoek" [Sweeping leprosy patients off streets, plans for isolation in the Sorok Island], *Chungang sinmun*, December 5, 1945; "Pang'im toen nahwanja Sorokto ro hosong" [Neglected leprosy patients, transported to Sorok Island], *Tonga ilbo*, December 10, 1945.

18 R. G. Cochrane, "Editorial—Sulphone Treatment of Leprosy," *International Journal of Leprosy* 17, no. 3 (1949): 299–304; Norman R. Sloan et al., "Sulfone Therapy in Leprosy: A Three Year Study," *International Journal of Leprosy* 18 no. 1 (1950): 1–9; and Luiz M. Baling and Raul P. Valdez, "Reflections on International Leprosy Congresses and Other Events in Research, Epidemiology, and Elimination of Leprosy," *International Journal of Leprosy* 62, no. 3 (1994): 412–427.

19 Jane Kim, "Leprosy in Korea," 114–179.

20 "P'ŭromisol nabyŏng e yuhyo" [Promisol, effective for treating leprosy], *Kajŏ ng sinmun*, July 12, 1946; and "Nabyŏng t'ŭkhyoyak 'P'ŭromin' ipko" ['Promin' especially effective for treating leprosy has been warehoused], *Tonga ilbo,* September 29, 1947.

21 Jane Kim, "Leprosy in Korea," 180–204.

22 "Leprosy," *Report of the WHO/UNKRA Health Planning Mission*, 55–59.

23 Nick Cullather, *The Hungry World: America's Cold War Battle Against Poverty in Asia* (Cambridge, MA: Harvard University Press, 2010); and David Ekbladh, *The Great American Mission: Modernization and the Construction of the American World Order* (Princeton, NJ: Princeton University Press, 2011).

24 Gavin Daws, *Holy Man: Father Damien of Molokai* (Honolulu: University of Hawai'i Press, 1984); US National Parks Service, "A Brief History of Kalaupapa," https://www.nps.gov/kala/learn/historyculture/a-brief-history-of-kalaupapa.htm.

25 "Director's Correspondence—Hansen's Division, 1952–53," RG 325–3–9, Papers of Board of Health of Hawaii, State Archive of Hawaii, Honolulu, Hawaii; and Sloan et al., "Sulfone Therapy."

26 Dharmendra, "Comment on Leprosy Control in Korea," *Taehan Na Hakhoeji* [Korean Bulletin of Leprosy] 4, no. 1 (1967): 119–123; Jane S.H. Kim, "Desegregate at All Cost!—The World Health Organization and the Battle for Leprosy Control in South Korea, 1966–1982" (paper presented at the Conference of the Association of Asian Studies in Asia, Taipei, June 23, 2015); National Human Rights Commission of the Republic of Korea, *Hansenin inkwôn silt'ae chosa* [Investigation on the conditions of human rights of the Hansen's disease patients] (Seoul: National Human Rights Council of Republic of Korea, 2005); and Anti-Corruption & Civil Rights Commission, "Hansenin kwŏnik kanghwa mit chngch'akch'on hwan'gyŏng kaesŏn pang'an" [Increase of the rights of the Hansen's disease patients and improvement of resettlement village environment] (Seoul: Bureau of Welfare and Labour Complaints, Anti-Corruption & Civil Rights Commission, June 2010).

27 Dharmendra, "Comment on Leprosy Control in Korea."

28 Han, Sang-t'ae, "WHO sajŏltan i pon Han'guk pogŏn saŏp—Maektonaldŭ pogosŏ rûl chungsim ŭro" [The public health projects in Korea as seen by

the WHO Mission—the MacDonald Report], *Pogŏn segye* [Public Health World] (April, 1960): 40–42.

29 "Kajang ch'ulse han haewoe kŭnmu Han'gug'in" [Koreans working overseas who reached the top], *Chugan Chosŏn*, February 10, 1994; "2 wôl 3 il ch'wi'imsik Han Sang-t'ae WHO sŏt'ae samu ch'ŏjang" [February 3rd, appointment ceremony, Han Sang-t'ae, regional director of the Western Pacific Regional Office of the WHO] *Ŭisa sinmun*, January 30, 1989; and "WHO sŏt'aep'yŏngyang samu ch'ŏjang e Han Sang-t'ae paksa p'isŏn" [Dr. Han Sang-t'ae, elected as director-general of the Western Pacific Regional Office of the WHO], *Ŭigye sinbo*, January 19, 1989.

30 WHO, "WHO in the Western Pacific: Founding of the Western Pacific Regional Office," http://www.wpro.who.int/about/in_brief/history/en; WHO, *Fifty Years of the WHO in the Western Pacific Region* (Manila: World Health Organization, 1998); WHO, *Polio Eradication in the Western Pacific Region* (Manila: World Health Organization, 2002); and WHO, "Director General—Dr. Lee Jong-wook: Biography," http://www.who.int/dg/lee/biography/en.

31 Han Sang-t'ae, "WHO sajŏltan," 40–42.

CHAPTER 6

From Ruin to Revival

Mobilizing the Body, Child Welfare, and the Hybrid Origins of Rehabilitative Medicine in South Korea, 1954–1961

John P. DiMoia

Published in 1978, Cho Se-hui's *The Dwarf* appeared as a novel after having been serialized, the first section issued in 1975.[1] The story of a physically disabled, diminutive male, the narrative establishes a parallel between the protagonist's economic circumstances, his urban home targeted for redevelopment in mid-1970s South Korea, and his personal status as a member of the disabled community, marginalized by society. Although the story is set more than two decades after the Korean War (1950–1953), it captures a time when physical disability was rendered as problematic. South Korea was in the process of negotiating its social welfare policies, even as the problem remained visible, sometimes linked to wartime injury, and later to industrial accidents and similar incidents.

Disability and Perceptions of Vulnerability

This chapter considers the origins of rehabilitative medicine in postwar South Korea, arguing that the process of adapting and accommodating international models of biomedicine, primarily American, followed a series of transitions from an initial focus on wartime injuries through rehabilitation and the use of prostheses, shifting the focus of such practice to a target demographic of young children and civilian patients by the mid- to late 1950s. The emphasis on this newly emerging demographic created a gap between the perceptions of two sets of participants. Korean doctors recognized the broader category of child welfare as an ongoing set of social problems by the early 1960s. In contrast, many of the American actors had already proclaimed the nascent field a success at about the same time, restricting their emphasis to the issue of physical rehabilitation for limbs, thereby failing to address the broader South Korean social context. The "new" rehabilitative medicine, a hybrid of existing practices and newer influences, reflects the social complexity of the early 1960s, "child welfare" referring to a much wider range of problems specific to the Korean context.[2]

Following the Korean War, the images generated to document events reached a high level, particularly within the American context, where the war needed to be justified in terms of its conjoined ideological and material aims. Along with strident anticommunism, the work of providing medical and infrastructural relief soon became one of the dominant themes shaping the dialogue concerning the impact of American intervention, even as the conflict was increasingly portrayed in international terms.[3] A growing literature concerning the Korean-American relationship touches on the later period of the war (1951–1953) and subsequent social relief efforts, noting that many common practices would be reshaped by the powerful intersection of interests between actors based in South Korea and the United States, the practice of adoption serving as a key example of this type of heavily mediated form of exchange.[4] Indeed, images of Korean children and refugees moving south toward Pusan and living in temporary camps have provided some of the most common backdrops framing the representative image of South Korea, even well after combat had ended.[5]

Moreover, the forms of medical intervention conducted during the war contributed toward reshaping the practice of South Korean academic and military medicine, and indeed the broader South Korean medical community over the course of the next decade (1954–1961) would become one in which international models of medical practice and pedagogy were increasingly available as a resource for Korean medical practitioners seeking to refashion themselves.[6] Previously, German academic medicine of the second half of the nineteenth century, as mediated through the lens of Japanese colonialism (1910–1945) and its medical pedagogy, had been the core of biomedical training, and for this reason many scholars have rightly questioned the effects of this legacy.[7] This observation should not ignore the considerable legacy of traditional Korean medicine, or *hanŭihak*, as well as the practice of a diverse collection of shamans, drug-sellers, and others who pursued their livelihoods despite the constraints of colonial rule.[8]

Given the enormous significance of medicine in its diverse forms, what needs to be called into question is any narrative assuming its unproblematic character after 1945, most of the scholarship to date focusing on the colonial period and the conspicuous excesses of Japanese colonial rule. Physical anthropology, anthropometrics, and a rigorous policing for public health and epidemic disease (cholera, typhus) are among the legacies associated with the colonial project, leaving behind a difficult heritage, one that represents a relatively new form of inquiry for Korean studies, and indeed, for the history of medicine field.[9] At the same time, the emerging historiography associated with the formation of a new nation, South Korea, has tended to overlook the continuities and ruptures informing this medical legacy for the period of American occupation (1945–1948), or looking ahead, the Korean War and the period of postwar reconstruction. Because a significant proportion of biomedical personnel received Japanese training, and given the exigencies of wartime and military medicine, it should not be surprising that at least some aspects of a "Japanese" medicine continued to inform medical practice.

In addressing these issues, this chapter interrogates the rapidly changing image of a medical subfield, rehabilitative medicine (*chaehwal ŭihak*), immediately before, during, and following the Korean War, through its major sites of practice. The first of these, the National Rehabilitation Center (Kungnip chungang chaehwal ŭihagwŏn) in Tongnae (near Busan) represents the wartime and immediate postwar legacy of the field in South Korea—its facility designed for the care of injured soldiers taking on a reconfigured form in the mid-1950s and receiving financial assistance from the United Nations.[10] Along with the Tongnae site, the Institute of Physical Medicine and Rehabilitation at New York University (NYU) was one of the world's leading centers for rehabilitative medicine. Under the guidance of Howard Rusk, the clinic shifted focus to the South Korean context, an increasing number of Korean doctors traveling to and from the United States for access to the most recent forms of clinical pedagogy.[11]

By establishing a relationship between these two sites, Tongnae and NYU, and tracking an emerging economy of medical exchange between South Korea and the United States, we can begin to discern the patterns of exchange consistent with the reimagined, postwar relationship. Beginning with the American occupation, biomedicine was a critical part of the

Korean-American relationship, challenging the legacy of its existing forms, here referring to at least three distinct practices: traditional Korean medicine, Japanese academic medicine, and missionary medicine as part of the Western presence. At the same time, the relationship between Rusk and the Tongnae site should not be reduced to a simple binary, or a technology transfer narrative, because this approach would assume an unproblematic relationship of donor and recipient. The latter part of this chapter therefore introduces the work of Oh Chung Hee (1926–1995), a female physician who traveled and practiced rehabilitative medicine in both contexts, ultimately receiving credit as one of the major Korean actors shaping the field as a form of practice.[12] Oh was also part of the national Child Welfare Survey in the late 1950s (published in 1961), underscoring the need to place Korean rehabilitation within a broader context of social and health issues.

Setting the Korean War as its background, the chapter examines the growth of rehabilitative medicine as a field, arguing that a conspicuous gap existed between the Rusk viewpoint from abroad and the material circumstances in Korea, providing rehabilitative assistance to a wide range of new populations. This assistance took the form of a diverse array of activities, including providing prostheses, offering physical therapy, and giving psychological counseling consistent with reintegration into society. It also began to include "social disabilities" specific to the local context, such as the growing population of mixed-race children.[13] Many commentators have argued that the relatively slow adoption of rehabilitative practices into Korean society had a great deal to do with the national economy, especially during the lengthy period of reconstruction associated with President Syngman Rhee (1954–1960). Economics alone, however, cannot adequately address the dense cluster of social issues grouped under the umbrella of child welfare, and the 1961 survey indicated that Korean practitioners conceived of rehabilitation and disability in a way often distinct from that of their American partners, which was far more reduced in scope.

In other words, the Korean setting offers a way to address the differences between the eagerness with which Korean doctors first traveled to the United States to study in clinical programs (NYU), and the subsequent difficulties some of these same individuals experienced on returning home (Dr. Oh). The emergence of rehabilitative medicine carries with it a corresponding package of related developmental and cultural assumptions, aims requiring time and resources to accommodate.[14] Specifically, the recovering nation required time to adapt its baseline assumptions providing a

Korea-specific foundational basis for a program of rehabilitative medicine. Even as the Republic of Korea (ROK) managed to subsidize this type of medical care in the absence of a national health insurance scheme, a practice that did not appear until nearly a decade later (1963), the nation struggled to find a place for these many of these patients, such as the main character in *The Dwarf.* Rehabilitative medicine in South Korea preceded the material recovery of the mid- to late 1960s, mobilizing its recovering subjects as a form of symbolic transformation, whereas in fact prospects remained unpredictable.

From Occupation to War (1945–1950)

Most accounts relating to the role of the United States Army Military Government in Korea (USAMGIK) emphasize the improvised character of the occupation, especially the negative impact it would have on political developments.[15] Despite these accounts, considerable attention was devoted to public health and medical concerns prior to arrival, cumulative experience in previous theaters of war easing the adjustment of the American Tenth Army. As early as 1942, students enrolled in Civil Affairs Training School (CATS) programs based at the campuses of a number of major universities received instruction concerning the problems anticipated in a postwar environment.[16] One of the major figures associated with the growth of the public health field in the first half of the twentieth century, Dr. C.E.A. Winslow prepared a series of lectures for the Yale version of the CATS program, many of his students later traveling to Korea and Japan.[17]

This style of preparation should not be read to suggest that the United States was well prepared for the Korean context, and in fact, occupation forces frequently overlooked and ignored existing forms of practice.[18] Instead, it should be seen in terms of an ambitious, universalist ethic informing the program and its subsequent iterations in mid-century American foreign policy and area studies, demonstrating the naïve confidence that a generalist approach might accommodate any contingency.[19] Students training for a range of postings—including Germany, Austria, Japan, and Korea—took classes alongside one another, many of the core elements being characterized by a common curriculum. Only in areas such as language did the curriculum reflect the particular contexts into which these specialists would be deployed, demonstrating an expectation of adaptability, one granting leeway both to the individual and to the models of knowledge he or she carried to a new context.

CATS assumed that its approach could fit any setting, given enough time and resources to make adjustments. In terms of building a narrative for broader issues of health and medicine, USAMGIK sought to link itself within the history of missionary activity in Korea, using personnel with such experience whenever possible. Even if the connection between the US military and the pre-1945 activities of missionary families such as the Underwoods (Yonsei University) was not direct, the enrollment of such groups within occupation activities gave authorities a baseline knowledge of the local context and thus a claim to legitimacy.[20] This approach not only lent a greater degree of confidence, but also marked a conscious break with Japanese colonial rule, another key factor.

Rehabilitative Medicine in the American Context

The physical body and its social context have a long history in the conduct of war and its aftermath. In the American context, World War I is generally considered the catalyst driving new developments, soldiers injured in combat proving a challenge to surgeons.[21] Rehabilitative medicine, especially the widespread use of prosthetic limbs, has been treated similarly, works such as Beth Linker's *War's Waste: Rehabilitation in World War I America* (2011) providing new insights. According to Linker, a growing faith in the power of medicine along with a desire to limit the distribution of military pensions—a practice deriving from the Civil War—allowed twentieth-century American reformers to "rebuild" soldiers injured on the battlefield, making them useful to society. Tasks such as auto repair, a job useful to both the individual and the employer, would allow a patient to remain seated while working.

For the Korean context, this history offers a useful way of introducing some of the same issues concerning the integration of new fields of practice into a distinct cultural setting. At the same time, many of these issues did not translate as easily as American actors might have expected. Both the USAMGIK occupation and the Korean War brought transformations to the Korean medical academy, plastic surgery from the war being one example of an adaptation stemming from a short-term contingency.[22] Patients who suffered burn damage from napalm were among the first to receive treatment through skin grafts and pedicles (flaps of skin attached to grow new skin), allowing doctors to improve their technique.[23] For rehabilitative medicine, this type of argument serves as well, the injured receiving access to prosthetic limbs from abroad, frequently through the Rusk Institute. At the same time, putting these same individuals back to work and creating a corresponding system of social and psychological support required considerable additional resources, meaning that the occupational therapy portion of the cycle was not necessarily available in mid-1950s South Korea. In this context, the prospect of employment represented a future project, an ambition to be conveyed through the work performed on the body of the recovering veteran. Moreover, distinct populations had problems unique or specific to the Korean setting (mixed race, polio, orphans), issues that could not be addressed immediately.

As Linker explains, the American context transformed gradually from a mid-nineteenth century model under which a soldier had to provide evidence of a debilitating injury to receive a pension. This approach was typical after the Civil War, and cases could be quite contentious because the state was aware of the burden of caring for veterans. The move toward prosthetic limbs, and later toward rehabilitative medicine, was designed to keep these veterans active and "useful," both to themselves and to society. This trend culminated in the formation of the Veteran's Administration (VA) system of hospitals after World War I. For its part, South Korea encountered this cumulative package of medical practices and social planning, but did so at a time of enormous devastation under weak economic conditions. The culturally embedded assumptions associated with American rehabilitative medicine did not necessarily fit the Korean setting. The social part of the plan suffered greatly, therefore, and Korean actors had to selectively borrow and adapt new practices after 1954, driven by the concerns of local needs.

The Korean War (1950–1953) and Reconstruction (1954–1960)

In terms of rehabilitative medicine, the war provided a context in which to mobilize material and infrastructural aid in large quantities, American and international partners rushing to aid a "Free World" neighbor. The arrival of American forces saw the intervention of mobile hospitals and the arrival of medical assistance from partner nations other than the United States. Norway sent its NORMASH units to accompany its forces and Denmark provided a hospital ship, the *Jutlandia,* based off the coast of Pusan, a site removed from combat.[24]

The earliest references to a facility associated with rehabilitation—defined here in the broadest possible terms—associate the area around Pusan with such activity, a hospital for wounded Korean veterans (*Sangi-gunin chŏngyangwŏn*) opening at Tongnae in 1952. The critical point is recognizing not just the wartime context, but also the significance of South Korean military medicine (*kuninŭihak*). The conjoined circumstances of the occupation and the war meant that public health and healing practices of all kinds were increasingly out of the hands of local practitioners (*hanŭihak*) even as many of these individuals continued with their practice. More important, given the extent of conscription, and the practice continuing into peacetime, many Korean males were exposed to new public health ambitions, meaning that the ROK Army became one of the major patrons of biomedicine.

The circumstances under which this facility functioned remain unclear, given that most sources mention its founding but offer no details about its operations.[25] We should therefore assume a care facility for veterans and injured soldiers, and, given the wartime context, one informed not only by local concerns but also by the growing availability of international medicine and supplies. During the war, cases of Korean surgeons working alongside their partners in field hospitals were numerous, and it is likely that this facility reflected a shared ethic as part of a project of providing relief and transition toward peacetime.[26] Dr. Oh, for example, began interning with the American Eighth Army as early as 1952 (1952–1955), during the war, before emerging as a major figure afterward. Again, the influence of external forms of rehabilitative medicine before 1950 was minimal. It is equally difficult to say how much the field had an impact during the war.

Along with sites for wounded veterans, facilities for two related groups of displaced populations—referring here to orphans and war widows, groups forced to migrate—were also established. The former group was perhaps the most conspicuous, especially given the migration associated with the first year of the war (June 1950 to spring 1951), in which a large number of refugees made their way south to Pusan. Widows receive far less attention in the literature, but would later attract attention as a population in need of assistance. Wartime references to this group indicate attempts to provide shelter similar to that for wounded veterans; after the war, new programs emerged for providing them with social relief. Specifically, efforts were undertaken to provide retraining with an emphasis on domestic industry, given the recognized need for these individuals to be able to care for themselves and eventually to secure employment. From these humble wartime origins emerged the basis for a new postwar program in rehabilitative medicine. Further, many of these wartime groups would effectively disappear as a concern for newly arriving American actors, even though remaining of great interest to their Korean counterparts.

Howard Rusk

The individual credited for transforming the rehabilitative medicine community in the United States was Dr. Howard Rusk, affiliated with New York University and known for his previous work with injured American airmen. This group of veterans from World War II was Rusk's patient base and enabled him to develop his work with prosthetic limbs and physical medicine. Still, Rusk had encountered considerable resistance when seeking additional sources of funding, given that the period after World War II (1946–1950) faced budget constraints, and he was competing with the VA health-care apparatus for resources. This competition was particularly relevant to the outbreak of the Korean War as President Truman's popularity declined and the American national budget faced constraints.

Even with the outbreak of the Korean War, Rusk found his options somewhat limited during its first two years as Congress concentrated on combat rather than the possibility of offering relief or rehabilitation. In fact, Rusk would not demonstrate a personal interest in Korea until invited to join an inspection tour (1953) by the American-Korea Foundation (AKF), an organization whose members included many prominent political and business leaders, headed by Milton Eisenhower, brother to the president.[27] Aware of the opportunity to make site inspections part of the tour, Rusk recognized the possible fit between his ambitions with rehabilitation and the Korean context given the numbers of injured patients. Officially, Rusk would become part of AKF's project of providing relief in the aftermath of the war, but he had much to gain in return, especially in terms of locating a new research context for his work. In effect, Rusk needed the Korean setting because it provided bodies and human subjects as a context in which to further develop his work.

At the outbreak of the Korean War, Rusk had already been a prominent name within American medical circles, overseeing the opening of the Institute of Physical Medicine and Rehabilitation at NYU. Still, his decision to join forces with AKF transformed him rapidly into an international figure; he assumed presidency of AKF in 1954 and in 1955 took charge of the World Rehabilitation Fund. He would later work in the Taiwan and still later the Vietnam War contexts, continuing to expand the scope of his research. South Korea was therefore the beginning of a second and much more ambitious career, one he pursued for the remainder of his life, extending his work with artificial limbs and developing a new range of therapies to countries in need. In turn, he received the chance to work with a much wider client base, and his work received far greater exposure internationally. Within these limited terms, he could claim great success, often overlooking the more complex sets of social circumstances and assumptions tied to distinct national sites.

Mobilizing Postwar Medical Transformation (1954–)

The fit between the Korean context and Rusk's project appeared well matched at first: he had the ambition to develop the field of rehabilitative medicine, and here were thousands of potential candidates waiting for new limbs, along with an accompanying range of new therapies. Both materially and ideologically, rehabilitative medicine rested comfortably within a larger pattern of developing networks of Free World assistance, South Korea receiving aid from its partner nations, providing a satisfying testament to its apparent recovery. Moreover, the channels that Rusk would establish between NYU and South Korea were paralleled by similar patterns of medical exchange established immediately after the war (1953–1960). Prominent among these efforts would be the Minnesota Project (1954–1962), established between the University of Minnesota and Seoul National University; the provision of additional funding to Yonsei University / Severance Hospital through the China Medical Board; and the establishment of the National Medical Center (Kungnibŭiryowŏn), a Scandinavian effort jointly overseen by Norway, Denmark, and Sweden, to continue work started as part of the wartime effort (1958–1968).[28] For many young Korean doctors, Dr. Oh among them, these new sources of funding provided new networks and the opportunity to study and practice abroad. Oh, for example, took advantage of both NYU and Denmark in the form of overseas visiting residencies.

For the American context, the AKF mobilization fit nicely with this pattern of overlapping interests, matching new sources of funding with previous networks of donors, such as missionary activity dominating throughout much of the preceding late Chosŏn and colonial periods. The Korea mission could no longer be framed exclusively in Christian or religious terms and now began to reflect a much broader set of concerns, appealing to the emerging postwar language of humanitarian concerns but not neglecting a healthy dose of Christian anticommunism in Asia. This claim would be most conspicuous in the case of Yonsei University, given its close ties to the Luce family and the United Board for Christian Higher Education in Asia, not to mention the movement of China Medical Board money to other parts of Asia.[29] As for the AKF, a quick glance at the names listed on its publications reinforces this observation, among them prominent individuals from the Eastern seaboard, including not only Dr. Rusk but also members of the Rockefeller family, and of course, holding out the possibility of contact with the Eisenhower family.[30]

In material terms, the significance of this AKF activity lies in the possibility of postwar study tours, the opportunity for Korean doctors and nurses to make their way abroad to work with new models of clinical practice. In their work on Korean science and technology networks, Moon Manyŏng and Kim Yŏngsik document that this possibility existed before 1945, but only in limited numbers.[31] For most pre-1945 students, the travel abroad option meant study in Japan or China, the former being the dominant model for elite students hoping to get ahead.[32] The transition to Western Europe and the United States came only with the Korean War and its aftermath, especially the corresponding growth of relief efforts.

For NYU, likewise, the arrival of Korean doctors in the second half of the 1950s was not a stand-alone project—even though clearly a publicity coup for the institute—in that these arrivals were among a wider stream of visitors drawing from populations around the developing world.[33] The work of rehabilitative medicine brought attention to the NYU site, and this interest was complemented by America's rise to the forefront as a major scientific center, supplanting Europe. Beginning in the mid- to late 1950s, a small number of Koreans began to arrive, most already holding medical degrees and at least some training or clinical experience.[34] In other words, these were not students seeking an overhaul of their medical education, but instead highly trained professionals seeking exposure to American models of practice before returning to their existing placements within South Korea.

Meeting Its South Korean Counterparts: The Institute of Physical Medicine and Rehabilitation, NYU (1955–)

These placements were based largely in the vicinity of the Seoul metropolitan area. As for gender, the first groups of trainees were largely male, though it is unclear whether this is because the institute chose to publicize male doctors more visibly in its publicity materials.[35] Similarly, the age range appearing in pamphlets suggest a marked emphasis on youth, doctors in their early thirties tending to predominate. The NYU group is similar to Korean medical personnel from this period, most completing their university education during the late stages of the colonial period, only to undergo a dramatic change with the transition to newer models deriving from international sources—the United States remaining the most favored destination.

Interestingly, at least two among this first group of Korean doctors, especially those associated with university hospitals, list their positions in terms of the field of physical medicine, or some close approximation.[36] These materials were assembled by NYU, so again it is not clear which party is responsible for translating the categories from Korean to English. Presumably these individuals had the option to self-report and represent themselves as they wished. In other words, it is difficult to know the precise material practice, but certainly the desire to engage with the new field existed at a number of Korean hospital facilities. Like Dr. Rusk, these doctors recognized the possibility of career growth in the postwar atmosphere, especially given the large number of patients and the possibility of financial subsidy.

These two major sites aside (specifically, Hanyang University and Woosuk University, along with their affiliated hospitals), however, Korean universities tended to link their developing interest in rehabilitation to one of two established fields, either pediatrics or orthopedic medicine. The former makes sense, because the motivation for assisting younger patients, those with the greatest potential for a long-term recovery, was particularly strong. As for orthopedics, the field was already well established, and interest in the musculoskeletal system remained strong both during and after the war, often a starting point for developing additional forms of medical specialization.[37] In both cases, the desire to link new knowledge to an existing context reflected a logical if conservative turn given that the long-term future of the new field remained uncertain.

NYU affirmed its long-term commitment by continuing to attract large numbers of international residents through the mid-1960s, Koreans prominent among this population. On average, eight to ten Korean doctors trained at the institute in a given year for at least a decade (1955 to the mid-1960s), indicating their willingness to take advantage of the opportunity to learn the skills associated with this new field. For example, Kim Ki Ho appeared in the institute's brochure, explicitly recognizing Dr. Rusk as the "father of Korean rehabilitation" and citing his prior context, the National Rehabilitation Center, as a "branch office," explicitly linking the two sites within a networked relationship.[38] This characterization reflects the goal of the participating American actors, the establishment of a network; the question remains as to what this activity meant for the potential transformation of the Korean site. Again, the presence of actors such as Dr. Oh Chung Hee suggests a more complicated story, one challenging the smoothness of this "successful" network story.

Tongnae (1954–)

The South Korean context included not only the financial issue of gathering support for rehabilitative medicine, but also the situation of adapting the field's concerns into a set of pragmatic and manageable goals fitting a new context. The site that became the National Rehabilitation Center held a set of associations, not all of which necessarily suited the program Dr. Rusk promoted. Because the Korean context left behind its practice deriving from wartime, such as providing shelter for orphans, widows, and wounded veterans in a series of "receiving homes," the notion of rehabilitation came to suggest a set of aims involving more than maintaining the care and physical welfare of the individual. The newer programs sought to foster recovery of the body and the spirit as well as to provide for an opportunity to become a member of society through job training.

In terms of institutional support, the NRC had a number of patrons with distinct agendas, making it a well-funded if not always legible institution. The infrastructure as of 1953 included a nine-hundred-bed facility but, according to American visitors at the time, never fulfilled its original intent as a rehabilitation center.[39] Plans to reconfigure the site as the NRC then shifted in emphasis to military veterans, especially those who had "reached their maximum hospital improvement," leaving additional room for rehabilitation training.[40] The motivation for such a gesture, seemingly directed at the enhancement of life for these individuals, was pragmatic, freeing up hospital beds for the Korean military. Along with the Ministry of Social Affairs, site sponsors included UNKRA, the AKF, and United Nations Civil Assistance Corps in Korea, meaning a great deal of resources, but also the potential for confusion over coordinating its aims.

In its publicity materials taken from the Tongnae site (1954), UNKRA depicted a wide range of therapies, including images of both children and adults undergoing a variety of rehabilitative exercises. In one image, children lie on the floor, working with adult therapists, smiling as they stretch their bodies on the padded surface. Circulated widely after the Korean

War, this style of image emphasized testifying to the effective use of funds. Similarly, an image taken from the same series offers a male and female patient sitting in close proximity, the male using some sort of pulley device to exercise his arms and upper body. We learn from the caption that neither patient was injured in the war. These are in fact civilian patients, the male suffering from severe arthritis. Even as this image offers a contrast to the previous one, providing content about specific types of exercises, the intent again is more rhetorical than content focused, offering a gesture of good will attesting to a recovering nation.

According to UNKRA materials, the appearance of a limited number of civilian patients by the mid-1950s was not at all unusual but instead indicative of a growing transition. As of 1956, the NRC housed approximately two hundred military subjects stemming from the war along with an additional group of slightly more than fifty civilians, about 20 percent of the total patient population.[41] Moreover, many of the civilians were young children, and the images offered by UNKRA attest to this trend as well, the accompanying images providing evidence of infants undergoing therapies of various kinds. A donation made by a charitable women's fund provided for the construction of an additional four buildings to the site, this new construction devoted to the care and welfare of children specifically, emphasizing a change underlying the founding principles of the site.[42]

This last statement should not suggest incompatibility between the care of wounded veterans and that of children, especially an orphan population; in fact a certain logic applied to consolidating available resources at Tongnae. Still, the UNKRA materials issued to publicize the newly opened NRC often underscore a shift in its basic orientation, one toward a younger and increasingly civilian demographic by the mid- to late 1950s. To review, early in the war, the site was generally associated with previous centers established for wounded veterans, and this activity continued into the early postwar period. In contrast, the NRC by 1956 was attracting attention for its ability to provide the latest forms of care and therapy, these tropes serving as an extended metaphor for the Korean nation. UNKRA specifically mobilized its language, using "from ruin to recovery" for its programs and the healing bodies of patients to testify to effectiveness.

From its modest beginnings, the NRC represented a close partnership of UNKRA, the ROK Ministry of Social Affairs, and the AKF, all three actors offering their collective funds, expertise, and materials. An estimated fund of more than half a million US dollars had gone into the planning and construction stages by this point, and additional donations continued to come in. Moreover, given an estimated fifteen thousand amputees nationwide, the site held a workshop for the manufacture of its own prostheses, meaning that the devices no longer had to be imported from abroad.[43] Along with living and therapy facilities, the center offered a range of possible options, claiming at least fourteen types of vocational training, indicating an embrace of the social aspects of rehabilitation. Even with its ambitions, this last claim raises questions, given not only the lack of resources but also the meager prospects for employment on the other end. In brief, the site carried with it the burden of an ambitious new medical scheme, as well as related social and ideological ambitions associated with a scheme for industrial recovery.

To summarize the series of transitions, the NRC transformed within its first decade from a wartime contingency into a major vehicle of positive publicity for its Syngman Rhee, South Korea, and its international patrons. For Americans in particular, it functioned as an ongoing testing site for newer therapies and materials, the target demographic taking on a much younger and increasingly civilian population. Again, this does not mean that the military origins of the site and its subsequent civilian usage were incompatible, but does suggest the increasing complexity with which the site was perceived, a variety of conflicting interests holding a stake in its outcomes. For the immediate postwar period, the site was concerned far less with providing social services than with creating a model space attesting to the successful transition away from the style of deprivation associated with the war.

Ongoing Negotiations: Oh Chung Hee at the National Rehabilitation Center (1952—1969)

The conjoined issues of mobilizing limited resources and translating international medical models to better fit the needs of the Korean context were common to Korean medical personnel of this period. Dr. Oh Chung Hee (1926–1995) not only experienced the challenges of new forms of training in at least two distinct overseas settings (New York and Copenhagen), but also rapidly became one of the central figures associated with the emerging rehabilitation center and is now recognized as one of the most influential figures in shaping the concerns of the field more broadly.[44] Initially trained in the Korean context, Dr. Oh spent a year with the Rusk Institute as a visiting resident (1956–1957), and later an additional year in Copenhagen (1963–1964), gaining access to European approaches. After returning to Seoul in 1957, she held a key position at the National Rehabilitation Center for more than a decade (1957–1969), giving her one of the longest tenures at the site as well as corresponding knowledge of its limitations and possibilities. Earlier, Oh had begun her education at Ewha during the late stages of the Pacific War, completing her medical education at Seoul Women's Medical College (1948–1951), bridging the occupation and the start of the Korean War. In terms of her unique approach to the field, Oh formed a critical relationship when she interned at the US Eighth Army Hospital (1952–1955), a choice that situated her within the wartime origins of the Tongnae site, but equally positioned her to help reshape its priorities as her career developed. As a Korean actor, Oh's public speeches and publications are quietly revealing in that they often exhibit her concern for the lack of resources and materials, meaning that she was not blind to the problems of the site and was eager to request additional assistance from her counterparts. As a medical doctor, moreover, she was critical of the Korean context when she felt that it failed to treat her work with sufficient urgency, again, placing her in a unique position as an actor with a range of views.[45]

This positioning brought with it, however, a comparable set of problems, in that Oh had to be responsive to numerous patrons over the course of her career. Her overseas study came with the expectation that she would represent the interests of her sponsors, at least in some sense. She also appears in a number of English-language publications as a symbol of hope for Korea. Her NYU venture was sponsored by the AKF, for example, and those who put her forward for funding were proud to claim their relationship, as was true for Anna Scott of Yonsei University.[46] For her part, Oh never publicly complained about the possible constraints of such a role, keeping her remarks to the problems of conducting new work in a field setting without attending to the circumstances. Implicitly, these types of critiques appeared in her remarks as a representative of the rehab center, suggesting that she knew the extent to which she could push her concerns without risk of offense.

Of the venues in which she appeared in this fashion, two are particularly worthy of note, a 1958 conference held in Australia and the 1961 national survey in which her chapter appeared, along with the work of many other prominent figures. In the first, which was published as an edited volume, *Conquering Physical Handicaps* (1958), Oh calls for pushing beyond theory, emphasizing the need for clinically oriented, practical experience.[47] Although the target of her remarks remains unspecified, she clearly urges a more flexible, hands-on approach to her field, as opposed to a static model drawn up in the abstract. Physical rehabilitation, framed in these terms, involved both daily struggle and incremental progress, given that it required a context in which real problems and issues must be addressed. In this sense, Oh indicates an awareness of her field as embedded within a larger social context, certainly the case for her participation in the 1961 survey.

This second volume captures the enormous complexity of Korean child welfare at the beginning of the 1960s; new sites and institutions starting to appear in conjunction with major hospitals meant that the National Rehabilitation Center now had a greater number of colleagues with which it might work. Yonsei University, for example, offered a Crippled Children's Center, combining the new impulse with its existing resources, some of these deriving from missionary and earlier activities. For her part, Oh contributed a chapter on physical rehabilitation, and its content is not radically distinct from that offered in her 1958 publication. However, whereas the former case was presented in Australia on a panel held in tandem with Dr. Rusk, the 1961 effort appears in a much more complex setting, as part of an effort to deal with the broader landscape of social welfare issues. In this second case, Oh had to situate her concerns within a Korean context that required greater care and an awareness of "disability," a category now holding a set of widely disparate meanings, though gathered as a cluster for the sake of clarity.

"From Ruin to Recovery": Rehabilitative Medicine in Transition (1961–early 1970s)

The recent literature concerning rehabilitation falls within the broader category of disability studies, and here the framework often invokes the language of citizenship to explain the process of gaining empowerment and reentering a community, regardless of perceptions concerning one's physical differences.[48] This citizenship trope is useful in the Korean context, even as we have to recognize differences across national lines, in that it recognizes how Korean War veterans were simultaneously valorized and marginalized after 1954. Their bodies proved extremely useful as a test case for a new biomedical field, and equally as iconic images circulated at the international level, celebrating the economic and political recovery of South Korean sovereignty. In this sense, rehabilitative medicine, with its assumption of transformation, offered a form of hope to Koreans, even while lacking some of the support mechanisms necessary to adapt the field to the new setting. For both soldiers and children, succeeding decades brought extensive negotiations to move away from older notions of the "sick body," and to construct the category of the disabled individual (*changaein*), one with a hindrance or problem to be treated, while recognizing the individual on their own terms.

In South Korea, one of the key gestures informing the provision of social services for these populations in transition involved the creation of a national health insurance system, which began only with changes to the basic health law in 1963 and ultimately started to resemble its present form by about 1977. Unlike celebratory accounts, such as Joseph Wong's *Healthy Democracies: Welfare Policies in Taiwan and South Korea* (2006), I argue that the South Korean health system of the military period cannot be read as progressive in its motivations, even with its expansion of care during the transition to democracy (1977–1989).[49] I am more interested in targeting its careful, selective exclusions than in the wider provision of care, and in the increasing trend toward privatization within the last decade. To circle this discussion back to rehabilitation, I argue that the National Rehabilitation Center, in its reformulation of the Tongnae facilities (1955–1958), focused initially on the wartime origins of the emerging field of rehabilitative medicine, especially its mission of providing a source of housing to veterans, orphans, and widows, and also replacement limbs. At the same time, this meant a more complicated relationship to a larger, and more diverse, population of the disabled.

As reconfigured by 1958, the Tongnae site attested to a nation on the road to recovery, one capable of providing employment opportunities for those with disabilities. The target demographic, the war having just concluded, reflected a shift to either congenital cases, or those with damage deriving from the symptoms of chronic disease. Most important, these newer patients were younger and civilian, and the emphasis was on recovery and integration back into family and work environments. This summary should not be taken as a skeptical reading of the motives underlying the material transformation of the site, but instead as recognition of the conspicuous differences between the American and South Korean contexts and their distinct sources of practice. In particular, a closer look at the provision of military medical care following demobilization would provide a much-needed supplement to the argument sketched out here, and for now, the key is to see that the ROK state envisioned a particular target demographic as recipient of its care. Military subjects were no longer as conspicuous or particularly desirable as these subjects by the late 1950s.

Assuming the previous statement holds true, where would rehabilitative medicine redirect its considerable energies? It is possible to see its imprint on the development of a diverse cluster of related fields within present-day South Korean medicine, including sports medicine, and—perhaps much more prominent—disability studies. In other words, rehabilitation, once removed from the immediate context of wartime contingency, began to focus its attention on other types of injuries, whether congenital, developmental, or owing to particular circumstances, such as the industrial workplace, another frequent site of attention in public health. Moreover, the new programs that would take root at major universities continue to lack an awareness of the wartime context, with Yonsei University (early 1960s) and Seoul National University citing later points of origin for their respective programs.[50] The point, again, is not to critique the historiography, but to recognize the vexed, contingent character with which rehabilitative medicine came to South Korea, making it relatively easy—and perhaps even more comfortable—to move away from the pain and suffering associated with the war, let alone the contentious colonial period.

Notes

Earlier versions of this chapter were presented at Binghamton University (2012), Johns Hopkins University (2012), AAHM (Emory University, 2013; UCLA, 2018), EASTS (Seoul National University, 2012), and Georgetown University (2014).

1 Cho Se-Hui, *The Dwarf*, trans. Bruce Fulton and Ju-Chan Fulton (Honolulu: University of Hawai'i Press, 2015). Thanks to Stephen Epstein for the suggestion.

2 Korea Child Welfare Committee, Han Welfare Committee, *Handicapped Children's Survey Report, Korea, 1961* [Han'guk changhae adong chosa pogosŏ] (Seoul: Pogŏn Sahoebu Han'guk Adong Pongni Wiwŏnhoe, 1961).

3 Howard A. Rusk Papers, Folder 288, "Report of Rusk Mission to Korea" (March 11–18, 1953), Western Historical Manuscripts Collection (WHMC), University of Missouri-Columbia; see in particular "Korea's Health Problems," 3–5.

4 Eleanna Kim, *Adopted Territory: Transnational Korean Adoptees and the Politics of Belonging* (Durham, NC: Duke University Press, 2010); Arrisa Oh, *To Save the Children of Korea: The Cold War Origins of International Adoption* (Stanford, CA: Stanford University Press, 2015); and see also

Bertha Holt (as told to David Wisner), *The Seed from the East* (Portland, OR: Holt International Children's Services, 1956). The growing literature on adoption notes a correlation with different forms of disability, especially true for mixed-race children, who were often targeted as subjects for international adoption. See also the *Journal of American-East Asian Relations* (24, nos. 2–3, 2017) for two special issues concerning the war and 1950s Korea.

5 "Report of Rusk Mission to Korea" (March 11–18, 1953), see "Korea's Child Welfare Problems," 5–8.

6 Hoi-eun Kim, *Doctors of Empire: Medical and Cultural Encounters Between Imperial Germany and Meiji Japan* (Toronto: University of Toronto Press, 2014).

7 Ibid. See also Jin-kyung Park, *Corporeal Colonialism: Medicine, Reproduction, and Race in Colonial Korea* (PhD diss., University of Illinois at Urbana-Champaign, 2008).

8 Soyoung Suh, *Korean Medicine between the Local and the Universal, 1600–1945* (PhD diss., University of California at Los Angeles, 2006).

9 On the question of physical anthropology and metrics, especially in the recent work of Kim Ok-Joo, see the *Korean Journal of the History of Medicine*. Hoi-Eun Kim has also presented on physical anthropology and published in the *Journal of Korean Studies*. For recent scholarship on mixed race and genetics in the Korean context, see Hyun Jaewhan, "The Making of a Genetic Nation: Human Heredity, National Identity, and Transnational Scientific Exchange in South Korea, 1926–2009" (PhD diss., Seoul National University, 2018 [Korean]).

10 *The National Rehabilitation Center, Tongnae* (Seoul: ROK Ministry of Social Affairs, 1958). The publication of this pamphlet was cosponsored by the United Nations Korea Reconstruction Agency (UNKRA) and the Ministry of Social Affairs. For UNKRA materials, see the United Nations Archives (New York), https://archives.un.org/content/united-nations-korean-reconstruction-agency-unkra. See also the Theodore R. Conant Collection at Columbia University, "A Tale of Three Beggar Boys—The Story of the *UNKRA Tongnae Rehabilitation* Centre." See also "Rehabilitation Programme—Physical Rehabilitation—Amputee and Prosthetics Reports," United Nations Archives.

11 Both the Rusk Institute (formerly Institute of Physical Medicine and Rehabilitation) and Bobst Library at NYU hold papers for this medical exchange. See *Records of the NYU Medical Center, 1942–1979,* Bobst Library, University Archives, New York University / RG 13. See also Howard Rusk, *A World to Care For* (New York: Random House, 1972).

12 Oh Chung Hee, Korean Medical Association Hall of Fame, Public Affairs (Kongjuksahwang), http://kams.or.kr/business/fame/sub1.php.
13 Ibid. See also works by Inga Kim Diderich (mixed race, social disability) and Hyun Jaehwan (genetics, physical anthropology).
14 John P. DiMoia, "From Minnesota to Seoul? The DeWall Helix Bubble Oxygenator and Technology Transfer in Open-Heart Surgery, 1955–1965," *Comparative Technology Transfer and Society* 7, no. 2 (August 2009): 201–225.
15 Allen Millett, *The War for Korea, 1945–1950: A House Burning* (Lawrence: University Press of Kansas, 2005).
16 Charles-Edward Amory Winslow Papers, MS, 749, Sterling Memorial Library, Yale University. See Series IV, "Academic Papers," Boxes 108–114 for CATS.
17 Records of the Army Specialized Training Division, Yale University, 1940–1963. See Accession 19ND-A-172, Civilian Affairs Training School. William R. Willard, first head of the USAMGIK Bureau of Public Health and Welfare, attended Yale's program.
18 For an overview of rehabilitative medicine in Korea prior to 1945, see Eunjung Kim, "History of Disability: Korea," in *Encyclopedia of Disability,* vol. 3, edited by Gary Albrecht (Thousand Oaks, CA: SAGE, 2005), 858–864.
19 Ada Leeke, *When Americans Came to Korea* (Freeman, SD: Pine Hill Press, 1991).
20 Horace Underwood, *Korea in War, Revolution and Peace: The Recollections of Horace G. Underwood,* ed. Michael J. Devine (Seoul: Yonsei University Press, 2001).
21 Harold Gilles and D. R. Millard Jr., *The Principles and Art of Plastic Surgery,* 2 vols. (Boston, MA: Little, Brown, 1957).
22 D. Ralph Millard, "Oriental Peregrinations," *Plastic and Reconstructive Surgery* 16, no. 5 (November 1955); see also Beth Linker, *War's Waste: Rehabilitation in World War I America* (Chicago: University of Chicago Press, 2011).
23 Millard, *Oriental Peregrinations*
24 The National Medical Center would later be formed from these tentative beginnings. See Jon Bjãr, *The National Medical Center in Korea: A Scandinavian Contribution to Medical Training and Health Development, 1958–1968* (Oslo: Universitetsforlaget, 1971). For the UNKRA context to this project, *see UNKRA—National Medical Center Agreement, 13 March 1956,* United Nations Archives.
25 James N. Calway, "Prosthesis for Korean Soldier Amputees," U.S. Army Forces Far East, Medical Section 1953 Sep. For missionary activity devoted

to amputees during the war, see also the papers of Reuben Archer Torrey, 1942–1966, Wheaton College.

26 Thomas Turner, "Chapter XVIII: Japan and Korea," in *Civil Affairs/Military Government Public Health Activities,* edited by John Lada, vol. 8 of *Preventive Medicine in World War II* (Washington, DC: Office of the Surgeon General, Department of the Army, 1976), 659–707.

27 Howard A. Rusk Papers, Folder 287.

28 Bjăr, *National Medical Center in Korea.*

29 The United Board for Christian Higher Education captures the transformation of Severance Hospital in its annual reports.

30 Howard A. Rusk Papers, Folder 287.

31 Moon Manyŏng and Kim Yŏngsik, *Han'guk kŭndae kwahak hyŏngsŏng kwajŏng charyo* [Documentary history of the development of modern science in Korea] (Seoul: Seoul National University Press, 2004).

32 Ibid.

33 "List of Foreign Residents," *Records of the NYU Medical Center, 1942–1979,* RG13.

34 Ibid. See also *Institute of Physical Medicine and Rehabilitation* (pamphlet).

35 *Institute of Physical Medicine and Rehabilitation* (pamphlet).

36 Ibid.

37 Seoul National University, for example, had a General Surgery Department before developing specialties in the early to mid-1960s.

38 Ibid.

39 These remarks appear in captions associated with UNKRA photos. United Nations Photo, "UNKRA-Sponsored Amputee Centre Brings New Hope to Crippled Koreans," https://www.unmultimedia.org/s/photo/detail/188/0188648.html.

40 *National Rehabilitation Center, Tongnae,* 9. The NRC was not a hospital, as the healing process was "complete," and the center was designed for rehabilitation only.

41 "Statistics," *National Rehabilitation Center, Tongnae,* 20–24.

42 Ibid., "Scope of Services," 10. See also "Tongnae Korean Physical Rehabilitation Centre—Photographs 4324, 4325, 4331," and "Tongnae Rehabilitation Center—Photograph 1803, 1804, 1805, 1806, 1809, 1810, 1811, 1812, 1813, 1814, 1815, 1816, 1817, 1818, 1819, 1820, 1821, 1822, 1823, 1824, 1937, 1938," United Nations Archives.

43 This was a goal at the NRC and during the war. See Hung Tong Chun, "Prosthetic Services in Korea," *Orthopedic and Prosthetic Appliance Journal* (June 1959): 85–90. Dr. Oh and Chun knew each other and worked together, and Oh cites Chun's work.

44 Oh Chung Hee, Korean Medical Association Hall of Fame.

45 Australian Advisory Council for the Physically Handicapped, *Conquering Physical Handicaps: Official Proceedings of the First Pan-Pacific Rehabilitation Conference held in Sydney, Nov. 10–14, 1958* (Sydney: Australian Advisory Council for the Physically Handicapped, 1958).

46 Anna B. Scott and her husband Kenneth M. Scott both worked at Yonsei / Severance, and along with Oh, published chapters in the 1961 survey as part of the Child Welfare Committee. Oh's chapter is titled "Physiotherapy of Children with Physical Handicaps."

47 Oh's paper in the *Conquering Physical Handicaps* volume is "Rehabilitation in Korea," 562–563.

48 Nancy J. Hirschman and Beth Linker, eds., *Civil Disabilities: Citizenship, Membership, and Belonging* (Philadelphia: University of Pennsylvania Press, 2015). For the Korean context, see Eunjung Kim, *Curative Violence.*

49 Joseph Wong, *Healthy Democracies: Welfare Politics in Taiwan and South Korea* (Ithaca, NY: Cornell University Press, 2004).

50 Tai-Ryoon Han, "Physical Medicine and Rehabilitation in South Korea: Past, Present, and Future," *American Academy of Physical Medicine and Rehabilitation* 5, no. 11 (November 2013): 964–969.

CHAPTER 7

Suffering Longevity

Life, Time, Money, and the Stem Cell Business in the Centenarian Era

Jieun Lee

> "BioCare Technology: Technology for Storing the Youth for the Healthy Future"
>
> "BioInsurance: The Wisest Preparation for the Healthiest Future"

The BioInsurance website in 2012 presented on its home page two phrases and two interactive images that effectively captured specific imaginations about the future that bear attention now. One image was an illustration of a double helix and the other a photograph of a happy, apparently middle-class family enjoying a holiday at the beach. The helix incorporated five clickable hexagons that revealed the names of different kinds of cells indicating different ways to "store youth." The message was simple. The cutting-edge technology to store stem cells isolated from different body parts would secure a person's healthy and happy future against possible adverse future health events. Although what is stored are stem cells, the catch phrase read "storing the youth"—not simply stem cells. The services marketed are "biocare" and "bioinsurance," that is, biotechnological care and insurance. This service is not care or insurance per se, however, but instead storage for stem cells or tissues that contain stem cells, such as placenta and cord blood. It thus offers the possibility for medical care in the future by ensuring that a person will have access to their own youthful stem cells. Put differently, the promise is about a chance to have a chance, an opportunity to have an opportunity when stem cell treatment turns out to be effective for a condition that might make a person ill.

This chapter takes the language used to market an emergent autologous stem cell banking service often called bio-insurance as an ethnographic entry to look into a kind of anxiety about the future and aging in contemporary South Korea (Korea).[1] Due to its novelty, however, the service still needs to create the market. It demands that bio-insurance companies develop a particular language to capture the potential clients' attention and persuade them to buy into the prospect of stem cell banking as a promising solution to the problems they face. Hence the marketing language needs to be crafted in a way that makes sense to ordinary people addressing what matters to them. It involves a certain work of "translation." Sociologist Michel Callon proposes that we attend to the process of translation in the constitution of scientific knowledge, as well as to how researchers position them as an obligatory passage point as they identify diverse actors and their problems in relation to their research program.[2] Marketing bio-insurance to people involves a similar process: they identify the potential clients, craft languages that touch on the issues that these potential clients might be concerned with, and relate those concerns to the promise of stem cell science. Yet the language crafted by those companies might at least hint at what is conceived as an important concern for the potential clients for stem cell banking services to attract them to pay several thousand dollars for cryopreserving their own stem cells.

What interests me in this specific translation is that it, when pronouncing stem cell banking as a form of "insurance" in order to attract the attention of potential clients, foregrounds a certain sense of insecurity, anxiety, and desire prevalent in contemporary Korea. The languages deployed for bio-insurance marketing, because they are aimed at ordinary middle-age Koreans who are not necessarily technology savvy, are supposed to address their concerns and suit a certain kind of ethos. Stem cell banking companies deeming bio-insurance an apt language to attract contemporary Koreans, and cryopreservation of stem cells as a way to "store the future" not only explains the concept of bio-insurance but also translates people's desires through the idea of banking stem cells. Rather than focusing exclusively on the bio-insurance marketing per se, I analyze the adjacent discourses on the future of aging often presented alongside the promise of bio-insurance through the mass media as well as in their own business strategies. In doing so, I take the marketing language as an object to investigate the "cultural imaginary" of the future of an aging individual and a new configuration of life and time in contemporary Korea.[3]

Recent discussion of "biocapital" in science and technology studies has drawn attention to the imaginations of future, and promises of new biotechnology are crucial to understanding the contemporary bio-economy.[4] Also, bio-economy brings the value of biological materials into view, given that the technological reformulation of living processes enabled the biological materials to "yield" vitality and profits.[5] The notion of bio-value that Catherine Waldby proposes helps us pay attention to the specific materiality of tissues as a valuable resource. My premise is that the profitability of stem cells, or a certain portion of their bio-value in the promissory market, depends not simply on technological manipulation of biological processes, but also on discursive reformulation of biological processes and entities into marketable ones. Building on these studies, I approach bio-insurance marketing as a specific articulation of vitality and promise in contemporary Korea in the burgeoning market for "anti-aging." Rather than exclusively focusing on the promise of stem cell technology, I discuss how the marketing language of bio-insurance makes sense in contemporary Korea because it draws on various idioms from other domains. Bio-insurance is marketed as a way to prepare for the later years in the context of anti-aging, and it resonates with the social discourse on aging as a serious problem that requires individual preparation. The promise (or blackmail) of bio-insurance as it is marketed now in Korea, I argue, is conceived by the ethos of preparation for a self-reliant and youthful old age, and anxiety over aging as a degenerating process. Here, the present is posited as the not-yet of the future, and in the context of health care, individuals are interpellated as not-yet-ill subjects. Bio-insurance persuades its potential client to conceptualize the present as the right moment to take an action upon the promise of enjoying surplus vitality in the future.[6] Stem cell technology has found its niche, derivative market in Korea where the not-yet-ness of the present underscores the discourses on aging.

"You Must Be Imagining Your Future This Way . . ." (But You Are Wrong)

Bio-insurance service is being offered by several biotech companies working on stem cell–related therapeutics. This type of service offers the companies an opportunity to earn cash that could be used for research and development. Those companies promote their business not only in one-to-one marketing but also at conferences. Stem cell banking services are rarely advertised in the mass media; companies use other channels instead to reach their audience. For instance, CHA Medical Group, which emphasized its expertise in embryology and cryopreservation, launched a membership-based, luxurious anti-aging center in Gangnam, Seoul, and is promoting the bio-insurance program as one of its services. Another company, Korea Stem Cell Bank, which started as an autologous stem cell bank in 2005, started expanding its reach in 2009 by collaborating with a personal finance advising company to sell the bio-insurance program through its sales network. Whether they are sold alongside personalized premium health-care services or financial insurance programs, both sell the stem cell banking service as a kind of insurance, but in a novel way—as different from ordinary insurance programs.

Bio-insurance companies store clients' stem cells isolated from their tissues for an extended period. The costs of services and contract periods vary, but it usually takes several thousand dollars to store stem cells for twenty to thirty years. They are extracted from various tissues (fat, bone marrow, peripheral blood) to be kept in freezers in anticipation of future use when the client is affected by a condition that will be treated with them. Company websites and brochures feature long lists of diseases and conditions for which clinical trials with stem cell treatments are being conducted, that is, for which stem cells are anticipated to be useful for treatment in the near future.[7] Because stem cells are known to differentiate into various cell types, the list includes diverse conditions that are related to degeneration of cells—strokes, myocardial infarction, diabetes, and Parkinson's disease, to name a few. Although most are still in an experimental stage, the list of ongoing clinical trials is presented as evidence of future therapeutic development despite the lack of known outcomes. The companies emphasize that stem cell therapy is a new medical paradigm that will become applicable to many conditions.

As an incipient and novel business that is still not popular among the majority of Koreans, it strives to create a new market by drawing on various idioms from other domains and explicating its benefit intuitively and simply. The explication itself, however, is not simple. It urges a person to imagine their possible future and present within the timeline of technological progress and bodily decline. If bio-insurance is based on the anticipation of the future, or more precisely, anxiety and promise about the future, what the marketing does is channel the anxiety through stem cell promise. Then, it can be said that the promissory biocapital, in this case, can be set in motion when the future is problematized and translated into the idiom of biotechnological promises.

As bio-insurance companies phrase their services as a way to "prepare" (*chunbi*) for the "healthy future" by "storing one's youth" (*chŏlmŭm-ŭl pogwan*), stem cells are deemed as an object that contains the youth in itself, and further as an object of investment: "*Save the health* [*kŏn'gangŭl chŏch'uk*] of all your family now. Cell Banking is not an insurance for monetary compensation, but *true insurance* [*chinjŏnghan pohŏm*] for a healthy future."[8] In this marketing, storing stem cells is presented as synonymous with, if not identical to, "storing youth," "saving health," and "preparing for the future." But what does it mean to equate banking or storing of stem cells to "saving health" and even "saving healthy future"? Do stem cells, health, and healthy future mean the same thing? Can storing and saving be said to mean the same thing? Moreover, what does it mean to say stem cell banking is a kind of insurance? In other words, what kind of future is that in which all these phrases supposedly mean the same thing? Nothing in any of these phrases is strange if one accepts that stem cells would someday be used for treating diseases in the future. Yet it is interesting to look at how these words are used interchangeably in the advertisement: stem cells, health, and healthy future. Stem cells appear to be an entity that promises health in the future. Storing, banking, saving, and insurance—all of these have certain economic connotations. When the words for monetary activities are used to explain the necessity of these services, they imply certain value and currency that is incorporated in the entities called stem cells. Here, one is led to see the value of stem cells that is not necessarily monetary, but bio-valuable.

How these words are associated as if they were interchangeable might tell us how the business of stem cell banking, using material specificity of stem cells as a kind of general potential and the technology of cryopreservation, promotes a specific imaginary of life, time, and money. Here, bio-insurance companies tell stories of (un)desirable futures to get them interested in the promise of stem cell banking. In doing so, they not only explain what stem cell banking is, but also ask people to imagine possible futures. Indeed, to see the value of stem cells preserved in a freezer, one needs to first imagine a future in which the potential of stem cells to differentiate into other cells will actually be a personal benefit.

To cultivate interest in bio-insurance, marketing needs to make people imagine themselves as a promisee of the stem cell promise. It needs to persuade someone to be a client who recognizes the bio-insurance as a promise for "me" and for "my future." It is a work of interpellation that involves marketing because potential clients need to first recognize it as a promising investment. Then, it can be said that the language of bio-insurance tries to create a kind of "surplus health"[9] by rendering what was not included in the domain of health care as an object of concern while enlisting people as potential investors in the vitality of tissues and the promise of stem cell sciences. To market a bio-insurance program as a way to prepare for future diseases, it is necessary to enable potential clients to imagine their future in a specific way—to highlight the possibility of being affected by diseases that could be treated with stem cells in the future.

Salespeople would need to frame bio-insurance as preparation for the future, perhaps by alluding to other preparatory practices such as pension funds.[10] The salesperson would ask a potential client to imagine his or her future in a particular way. "You need to imagine your retirement years as luxurious, such as enjoying golf trips with your wife, as you move your money into pensions. If your health does not allow this, your retirement fund could be completely exhausted in health-care treatment." The premise here is that you, the potential customer, are already aware of the necessity of preparing for old age. The promise of bio-insurance is to secure an affluent and happy life that you will enjoy thanks to what you are saving now. But you cannot necessarily prevent illness. The salesperson calls "you" into this narrative. The story reminds you that you are already and supposedly envisioning your future in a specific way, but something is missing from that vision. "You are planning to enjoy your retirement with the pension fund. But if you're affected by dementia or cancer before the happy retirement life, you can't enjoy—your life will be no more than dying day by day. What do you think about it?" Not good, right? Good life requires both money and health; if you are saving money, it's time to consider "saving health"—helping ensure a healthy future.

The salespersons are instructed that promoting bio-insurance is a different task than selling ordinary, familiar insurance programs. Cell banking service is new, and people might think it is too early to invest in it. The question, the instruction says, is how to sell a different value and what the different value is. One would not want to spend money on an unknown service. What matters is to emphasize the value of health, which is not reducible to money. Salespeople are supposed to get potential clients out of the monetary metrics, to make them imagine retirement in relation to quality of life in the contrast between the healthy and the unhealthy body, and to bring them back in the monetary metrics to ask them to buy this service at a low cost relative to that of losing health in terms of a normative body. In showing a video clip of a patient affected by Buerger's disease, for instance, a salesperson would ask questions the answers to which seem obvious. "Do you think that this patient wanted to get affected?" "Can you sell one of your legs for $1,900? How about for $100,000 or $200,000? If you could keep your leg for $1,900, what would your choice be?" The question here is clearly rhetorical. You cannot put a price tag on your leg. The present investment of $1,900 might seem substantial, but the return is invaluable.

Targeting mainly middle-class men and women in their forties and fifties, the salesperson urges a potential client to imagine a future in which one has money but not health. The future is rendered as a disastrous one—you might have prepared for it with a pension, but because you could lose your health it is not complete. Bio-insurance is rendered as an investment plan to prepare for the uncertain future. Money reserved in a pension is a resource for you to fully enjoy your life in the future, but some things cannot be bought with it—your leg, your health, your vitality. You cannot be active and lively if you lose any of these. Without health as a capacity, your life is simply a matter of dying day by day. This talking strategy is to encourage individuals to be concerned with future health, not necessarily in terms of mortality, but instead in terms of the capacity to fully and actively enjoy life. It provokes an anxiety not about death, but about being chronically ill.

The image of having a degenerated or devitalized body for an extended period seems to be crucial to this talking strategy. When a chronically ill person is said to be "no more than dying day by day," it is not death they are anxious about. Instead, it is the surplus years during which they cannot live lively, hopefully, and happily. When the salesperson says that your pension funds will all be spent to treat your incurable conditions, you as a patient are rendered as a person whose life is meaninglessly extended. This kind of life is rendered, in this narrative, simply superfluous. Without stem cells, your life will be hopeless.

Be prepared—otherwise, your future will be merely suffering for an extended period. You are not yet ill, but you will be. Save your health with stem cells—they will turn into anything you need in the future and regenerate your degenerated body. The regenerative future is yet to come. The present is the not-yet of future, you anticipate the future to come. But it is through preparation that you can secure a good life that is to come.

Suffering Longevity: Prepare for Your Future in the Centenarian Era

Bio-insurance marketing promises its buyers a good life in the future. The biological capacity of stem cells, still experimental and only in laboratory, is rendered as the crux of this promise: people would be able to prepare against aging by storing their stem cells. Yet the narrative works not only by the promise but also by the ethos of preparing, and by imagination of the future characterized by bodily aging and degeneration, which will impair life to the extent that it appears to be no more than dying day by day, surviving without meaning or pleasure. The entire set of value assumptions about the value of health and meaning of life, as well as aging as a degenerative process, is at work. Further, the value of a leg (incalculable) and the immediate cost for bio-insurance ("only $1,900") is not only a rational calculation, but also an imperative to prepare for the future—the future that is obviously unfavorable, yet imminent, without preparation. Here the business of bio-insurance, as much as it is promissory, requires an anxious subject concerned with something that is yet to come. That a person is healthy now does not eliminate the possibility of soon becoming ill. The present state of a person's life does not secure anything in the future. Most people will probably live longer than they expect, and their body will definitely face the possibility of illness for an extended period. To make matters worse, no one will take care of another person unless they prepare for themselves.

The Korean government and popular media have announced the advent of a new era, the centenarian (*paeksesidae*). The discourse on the aging society has shifted subtly but significantly by framing the problem of an aging population as an imminent reality and a personal matter that requires individual preparation. In his 2011 new year speech, for instance, President Lee Myung-bak emphasized that the centenarian era (CE) would bring a revolutionary change in people's lives. Indeed, it was a call for a revolutionary change in our view of life to be prepared for the future.

> Now, we are facing the era in which we should live our life *anticipating a hundred years' life*. . . . We must, however, face the reality—after a person who has been employed for thirty years retires, she or he should live longer than that. . . . We ought to open up the windows of opportunity to people at each momentum and situation in one's life. We should also provide conditions in which one can *develop the capacity for one's self-reliance and happy life*.

Policy reports that followed also describe the centenarian era in a somewhat different way than policies for an aging society. The National Research Council for Economics, Humanities, and Social Sciences, in particular, drafted a report titled "Centenarian Era, How To Live Happily: How Prepared Are We for the Coming Centenarian Era?"[11] The researchers explicitly distinguish their study of the CE from other studies on low birth-aging problems, or aging society questions conducted in Korea and other countries as it, as a future-oriented project, is concerned not only with the elderly, but also with the entire population at different ages. The report, pointing out the limitation of previous policy studies on the aging society, proposes to include "the entire generations while mainly focusing on the *changnyŏnch'ŭng* [people in their forties and fifties]" as subjects that require intervention.[12] The focus of CE policy is not on supporting the elderly, the already aged population, but on preparing the not-yet-elderly for the new era in which they are supposed to live up to a hundred years. It goes hand in hand with neoliberal emphasis on self-reliance and responsibility.

Coupled with low birth rate, aging society discourse has politicized life itself within the framework of a welfare society. In this discourse, what mattered is how to manage the productive population among the entire population within the nation. Hence, aging society discourse concerns the reproduction of the nation via maintaining its productivity. It has served as a framework to deal with the increase of the older population and the decrease of the productive population, decrease in economic growth rate, increase of public cost to support the older population, decrease in taxes collected, and conflicts between different generations related to the issue of taking care of the older generation. The discourse is ultimately about population and centers on the productivity and sustainability of the nation-state within the framework of the welfare state. If it concerns the future, it is to balance the productive and unproductive population in the nation.

Although the CE discourse and the aging society discourse share a problematic aspect, the CE discourse explicitly foregrounds a temporal narrative for an individual. Rather than focusing on the rate of unproductive population, it draws attention to the increased life expectancy of the presently productive population. It is concerned with the duration of individual life after retirement (more precisely after an individual drops out of the productive domain). The CE discourse is not simply another population crisis discourse about society as a whole. It is a discourse of (potential) crisis in an individual's life. Although it addresses the crisis of the welfare system, it is not as much about the restructuring of the system as of the individual's vision of life. Foregrounding the figure of an individual who is aged (not productive, not earning an income, losing health), the CE discourse warns that most individuals would live up to a hundred years, that is, too many years without regular jobs or health. If not prepared for this coming era, a person would be damned to a lifeless life.

Rather than viewing the society at large as a unit that needs intervention, the CE discourse notes that a fundamental restructuring of individual lives is needed in light of the prolonged postretirement years. A person now needs to rethink their life strategy to prepare for old age. Although the aging society discourse and the centenarian era discourse both concern the national future, the latter is marked by its focus on the individual and individual life plan. One needs to be aware that living will become more costly if a person is not only not productive but also unhealthy. Hence the CE discourse emphasizes individual awareness and planning—managing financial resources and health to prepare for old age.

Centenarian era discourse has not entirely replaced aging society discourse. Instead, the idea on which CE discourse is based simply added another layer to the aging society discourse—capacity building and future planning for the not-yet-old population. The additional legal clauses on low birth rate and aging problems are explicit changes in the government view. The practical solution of providing resources, such as consulting services to people to prepare for the postretirement years, points to unpreparedness as a problem of the centenarian era. It emphasizes the importance of predicting and preparing for the future by the individual over changes in the social welfare system. At this point, it is not simply a discourse on how to manage the population size; instead, it is a question of the conduct of individuals as capable citizens. The consulting service to be provided is part of capacity building, not simply building capacity such as new skills to do something, but building capacity to manage existing resources, one's personal property—capacity to manage what one already has for an uncertain future.

The CE discourse could be effective because it addresses individuals as the subject of preparation. It is not simply a futuristic discourse, but also a discourse that effectively captures people's anxiety on aging as they observe the increase of old people living without appropriate care. Its power is in the way it evokes a paradox of longevity. Rather than an aging society, which presumes aging of the population at large as a problem, the CE discourse portrayed in the popular media emphasizes the two sides of prolonged life. Longevity, often expressed as "centenarian" or "living to a hundred," has been traditionally said to be the dream of humanity. The CE discourse does not dismiss this, but asserts that it can be achieved only with health and wealth. Ironically, perhaps, longevity is to be suffered—unless one is prepared. This discourse has been actively circulated and promoted by popular media, the financial industry, and the health industry in a manner that "incites anxieties and confusion among people who are entering old age and their families," as one Korean gerontologist has lamented.[13] Financial institutions and the health industry in particular could exploit this discourse by emphasizing that the retirement years need plans and preparations. Unlike the aging society discourse that addresses the future of the nation, the CE discourse addresses the future of individuals. If you are well prepared, you will enjoy old age. If what is at stake was the shared future of the nation-state in aging society discourse, the CE discourse—taking the crisis of welfare state for granted—advises individuals that their future could be different if they plan ahead.

In the centenarian era, we will see people suffering longevity: too little death, too long a life, and too little security—unless prepared. Life in retirement is a kind of surplus. To deal with it, a person should also prepare surplus from now on to avoid being parasitic and hopeless later. Scarcity of health and wealth in the future will certainly come: one should be prepared for it. Retirement is a time for consuming rather than producing. Productive years are the time to accumulate for the consumption years. The question is how to manage the resources within this time limit. Seen this way, the CE discourse is not simply a variation on aging society discourse. It is instead one on managing resources in relation to time's dis-abling effect on capacity.

Being Not-Yet-Ill in the Centenarian Era

The policy concern about public health in the CE scheme has changed in a way to emphasize prevention. If preventive medicine has targeted communicable diseases for the last few decades, in this CE context, it has turned toward health promotion and lifestyle changes to deal with chronic conditions, particularly among the aging and aged population.[14] Given that the extended life of an ill population is costly, and chronic disorders such as high blood pressure and metabolic syndromes are anticipated to beget more serious diseases and costs, public health policy should be concerned more with preventive medicine and early diagnosis of such conditions. By providing various ways for individuals to identify potential health issues through early diagnosis and urging them to manage their health in an appropriate way, the government is trying to reduce the costs that would be incurred when the not-yet-ill population actually becomes ill.

Not only the government is interested in the vision of preventive medicine. Indeed, many existing clinics and hospitals are also invested in promoting early diagnostic services to maximize profit by incorporating the not-yet ill population into their market. Some clinics actively practice traditional Chinese medicine's notion of *mibyŏng* and translate it into a temporal term.[15] "If you're in the *mibyŏng* in which 'one does not have any disease [*chilbyŏng*], but is not healthy either,' one's immune capacity is to decrease, and accordingly the probability to get affected by disease is to increase."[16] Here, *mibyŏng* appears to be a kind of illness that has no proper name. It is a state that needs to be examined and treated; it is not quite a disease yet, but it is an indicator of future disease. If a person does not change their lifestyle and take good care of their body, they will definitely contract a disease or illness in the future. In the absence of intervention or self-care, *mibyŏng* will unfold into actual disease or conditions that could have a severe effect as time passes. It is this temporality of not-yet that *mibyŏng* discourse in contemporary Korea creates.

Originating from ancient Chinese medicine, *mibyŏng* refers to the state of being neither healthy nor sick and emphasizes the virtue of constant care of self within a given environment. The recent surge of *mibyŏng* discussion in the popular media as well as in medical literature (particularly in Oriental medicine), coupled with the idea of life as an accumulation of risk factors, adds another temporal dimension. It is translated as a category to name the in-between itself—the "gray health zone." The preventive care that needs to be taken for *mibyŏng* today is saving bodily capacity that will be lost in time. It is related to a specific temporality linked to the body as an entity that degenerates over time because of the risks accumulated through life. A newly established anti-aging clinic explicitly links the concept of *mibyŏng* to aging by highlighting the possibility of adult diseases. If the ancient Chinese medicine emphasizes the process to which one needs to attend as body and environment is entangled by the notion of *mibyŏng*, its contemporary version emphasizes the risk factors to be accumulated in the body and to finally turn into disease. At this point, it assumes a linear progress of time—change your lifestyle, otherwise, your body continues to accumulate the bad stuff from your life, and *mibyŏng* will grow into an actual disease. Health itself emerges as a goal to be achieved. It is not so much the absence of disease as having vitality and vigor. Then it is the bodily capacity that matters.

To an individual, health is also taken as a capacity. It is a capacity to live a "normal" life—being productive and not dependent. Health is not defined by its negative, but as a goal to be achieved and a state that can be lost in time. Moreover, health here is not the normal state of being, but an aim to be achieved by monitoring and management. One cannot simply be called healthy due to the absence of identifiable disease. If one is not perfectly healthy, one needs to suspect a problem that could unfold in the future into an actual disease. Health, defined not simply as a state of absence of disease, is not something usual. It is instead a goal to be achieved by constant monitoring and self-care—it is an exceptional state, like disease. What is presented as a normal state in this *mibyŏng* discourse is being neither healthy nor ill at present, but anticipating diseases in the future—the state of not-yet-ill. You are supposed to be suspicious of your health. What you have perceived as a normal distress, weariness, or discomfort becomes an indicator of disease yet to come.

Be prepared. Get examined. Monitor and manage yourself because you are not yet diseased, but your usual discomfort says that your body is accumulating adverse impact and losing its capacity to cope with future problems. The rediscovery of *mibyŏng* discourse exemplifies how the medical industry tries to capitalize on the temporality of not-yet by provoking anxiety over the future among ordinary people. It is the not-yet-ness that is constantly foregrounded to capitalize on anxiety about the future. In a sense, it is this surplus temporality that becomes the crucial element in the CE. A person is not simply living in the present. A person is not healthy but not ill. It puts people in the anticipatory horizon and urges a constant preparation for the uncertain future as a form of life. A new life form, frozen stem cells, is promoted by bio-insurance marketers as a novel technological object for this form of life. Framing autologous stem cell banking service as bio-insurance, the marketers urge people to imagine the adverse future to be insured against by stem cell banking. When the salespersons compare bio-insurance with other types of financial preparation such as life insurance or pensions, audiences are assumed already subject to anxiety for the future. Cell banking is a biologized future preparation driven by concern about health that these salespersons of bio-insurance are trying to promote.

Selling Promise on Promise

As a way of preparation, bio-insurance translates the matter of health into biological potential that one would lose if not acted on right now. It problematizes aging as an inevitable loss of bodily capacity and the body as an aging thing in a linear fashion. It problematizes not only the future, but also time itself, for the body is constantly aging. It implies that cells will also get old, losing their capacity and health. Future contingency is central in both existing insurance programs and bio-insurance, but bio-insurance adds another layer. Insurance services attempt to generate anxiety about the future and staging the risk as something one should be prepared for. If anything is new in the language of bio-insurance, it is the value of the individual body and the quality of life, which are rendered priceless, which need to be protected from the threat that increases with age. Loss in health cannot be reduced to monetary value. If you are affected by a degenerative disease and lose your leg, how can you measure the value of your leg?

When the salesperson explains what would happen if stem cells are not stored and the individual later becomes ill, what one does in the present (not subscribing the service) is a regrettable decision, an unwise choice. The salesperson draws two pictures of the future for the potential client—the promissory future of regenerative medicine using stem cells for therapy for various diseases, and the unfavorable future of being affected by degenerative diseases for which stem cells need to be prepared. By presenting these two futures as imminent, bio-insurance marketing brings the future into the present and makes the present the not-yet future. Conceiving the present this way, a person is urged to take action immediately, or at least promptly. For the future is presented not as indefinite and thus open to all sorts of possibilities, but instead as definite, one in which a person loses their health, the action to be taken in the present is obvious. "Save" health before you lose it. Save youth before you lose it. Stem cell is a container of youth and health as you now have.

Because it promises to keep an opportunity open, it is a promise on a promise of regenerative medicine, or stem cell therapy. In this temporal scheme, one is expected to envision the present in relation to the future and the present as a time for preparation for the future. The future is brought in to render the present as a moment to sign the contract for bio-insurance. The salesperson might not make any definite promise. It is not even promised that the stem cells would become useful. It promises only that a person will have their stem cells stored in the bank in the future. What is sold is the hope for cure, the opportunity to have hope in the future. It is a selling of promise on promise.

To have hope with this particular cell banking service, you need to imagine the unbearable future first. It asks you to imagine yourself in a sick bed rather than assume that you will simply get old. Buying an insurance policy is to hedge against future risks. When a person gets sick, the insurance company promises, the costs would be covered. You never know whether you will be sick in the future, but you can be sure that your illness would cause a financial burden to you and your household. An insurance policy asks clients to translate their life circumstances in monetary terms, and is developed in a way that translates various possible life occasions into risk profiles whose economic "value" can be calculated. An insurance policy is not simply hedging against risk. Indeed, it constitutes risk in financial terms, accordingly a specific kind of future for which a person needs to prepare financially. Because various forms of care have been commercialized, not only medical care, but also care for patients and seniors, it would not come as a surprise for an insurance company to offer dementia insurance. Yet to be attracted by such a proposal, a person would need to already imagine their mother with dementia as a burden and contemplate whether the promised cost of coverage for the uncertain event is a promise worth buying. An insurance policy is not simply about the future risk. Indeed, it requires a type of subject who can translate life circumstances in cost and financial terms.

If conventional insurance programs translate this health problem into economic terms, bio-insurance promises that it will do what traditional insurance cannot by replenishing injured tissues. In a sense, bio-insurance requires a subject who links aging to degeneration of bodily capacity, which can to some extent be enveloped in an entity that can be frozen and stored. If bio-insurance is a plausible business plan, it is because stem cells that can be isolated from the bodily environment and stored without losing their capacity could be one way in which the biological clock can be suspended. If traditional life insurance is an instrument that works because life circumstance can be calculated in monetary terms, bio-insurance highlights that it cannot be calculated, but instead needs to be prepared by storing the vitality capable of restoring bodily function. If life insurance in general focuses on the money's capacity to help a person deal with life circumstances, bio-insurance brings back the cells' capacity to replenish the body.

Money comes in again in this picture as a capacity to ensure that an individual can access their vital tissues in the future.

Bio-insurance marketers explain that a person needs to bank their stem cells immediately because cells lose their capacity when a person becomes ill or even simply ages. "Healthy stem cells *work* better." Quoting research data that relate bodily aging and the capacity of stem cells to proliferate or differentiate, bio-insurance companies emphasize that it is better to store stem cells sooner than later. Here, bodily aging and loss of cellular capacity make the present the time to take action. Now is when you are healthiest and youngest. It is as if a person's body ages to lose its vitality in a linear fashion. Although the occurrence of disease depends on multiple variables, its probability generally increases over time as a person ages.

Although the linear progress of time typically leaves degenerative traces on the body, it is also expected to show real progress and development in science and technology. Banking stem cells is to use them for therapy in the future. Although many diseases are listed as potential candidates for stem cell treatment, most tests are still at an experimental stage; it is not certain that the same treatments will necessarily be available in the future. However, stem cell companies emphasize that ongoing research worldwide indicates future availability of stem cell treatments. Here, the present is the past of the future, the moment for preparation.

Saving Surplus Vitality for the Future Yet To Come

Stem cell promise depends on the promise of regenerative medicine. The promise of regenerative medicine uses the figures of patients affected by degenerative conditions. The examples most often brought forward are Christopher Reeve, Muhammad Ali, and Michael J. Fox. Hwang Woosuk also used the figure of Kang Won-rae, a famous Korean dancer and singer suffering from spinal cord injury. They are brought to the fore not only because they are celebrities and people are familiar with them but also because they are lively people whose lives have dramatically changed due to injuries or diseases. The promise of regenerative medicine is not necessarily a promise to make dying people live, but rather a promise to give a "normal" life to people who have to struggle with those devastating conditions while not dying. What these celebrities had to face is not simply "death" of a biological body, but an "unhealthy" body with which one needs to live for a while—a kind of "dis-abled" body whose survival is supposedly costly.

The surplus vitality in stored stem cells whose potential is to be realized in the future, in the marketing language of bio-insurance, promises what might be called a "surplus" quality life against superfluous life. What bio-insurance promises is not simply extension of life by treating some fatal diseases such as cancer. It instead promises quality of life by curing degenerative diseases that would dis-able one's body and make one dependent on care by others. Once the vitality of stem cells is stored in the bio-bank, a person will regain the capacity to fully enjoy their life as an active, healthy, capable, and self-reliant individual for which one should be preparing.

Bio-insurance business capitalizes on anticipation, specifically, to realize the potential of stem cells. Indeed, bio-insurance can promise unspecified healthy future precisely because of the temporal gap between the present in which stem cells' vitality stays as unspecified potential, and the future in which it may or may not be applicable to specific diseases. The promise is sold to people who are still healthy but anticipating degeneration of bodily capacity and degenerative diseases. If stem cells are seen as bio-valuable in this context, it is because of this not-yet-ness of stem cells' potential and people's health. The creation of not-yet-ness itself enables the companies to accumulate surplus value. Then, the subject of this marketing is also urged to see themselves as not-yet-ill persons whose future would be bettered by the not-yet-ness of stem cells. By constantly emphasizing the uncertain future, bio-insurance marketing aims to speak to a peculiar kind of subject, or interpellates a specific kind of subject—a not-yet-ill subject anxious about their future.

Of course, technologies of stem cell extraction and preservation are important infrastructure for this business. A similar business model can be found elsewhere. For instance, cord blood banking has been popularized in many countries as a kind of biological insurance for newborn babies. What the bio-insurance business in Korea highlights is a specific articulation of surplus vitality and promise of regenerative medicine with anxieties about the future through bodily degeneration and the modality of self-care through preparation. Bio-insurance people say that a person should bank their stem cells before they lose their capacity. Resonating with the discourse on aging, bio-insurance also asks people to prepare for the uncertain future with a promise of regenerative medicine. As I showed by comparing the CE discourse and the aging society discourse, it is again about the individual's future—not necessarily related to the productivity of body envisioned in the industrial mode of production, but instead related to the capacity to enjoy life. If stem cells are thought to have value, it is not simply because they have the biological potential to regenerate bodily tissues. One interesting site in which stem cells are valued is in the anti-aging business that constantly posits time as a force that dis-ables the body. Time here is not simply duration but filled with everyday circumstances through which the body will accumulate bad stuff. In addition to natural or normal aging, the body also accumulates harmful substances and stress. The body, as it ages, anticipates degeneration. Stem cell technology promises, by suspending the biological clock and maintaining the regenerative capacity, to restore bodily capacity, and fight against the dis-abling force of time.

Notes

Epigraph: Bio Insurance website, accessed December 1, 2012, http://www.bioinsurance.co.kr.

1 The materials analyzed in this chapter were gathered during the author's fieldwork in Korea in 2011. The materials include brochures and promotion websites of bio-insurance companies, a transcription of a promotion session given to the potential clients, and blog posts from bio-insurance salespeople. All empirical material presented here is written or spoken in Korean and translated by the author. I borrow the term "bio-insurance" from a brand name for one of those companies that provide stem cell banking. Although the company took it as its own brand name, other stem cell banking companies also use a similar term to refer to their service—"bio-*pohŏm*" (insurance), or "true life insurance."

2 Michel Callon, "Some Elements of a Sociology of Translation: Domestication of the Scallops and the Fishermen of St. Brieuc Bay," *Sociological Review* 32, no. 1, suppl. (1984): 196–233.

3 I borrowed the term "cultural imaginary" from the cultural historian Graham Dawson, who defined it as "those vast networks of interlinking discursive themes, images, motifs, and narrative forms that are publicly available within a culture at any one time, and articulate its psychic and social dimensions." Graham Dawson, *Soldier Heroes: British Adventure, Empire and the Imagining of Masculinities* (London: Routledge, 1994), 48.

4 Nik Brown, "Hope Against Hype: Accountability in Biopasts, Presents and Futures," *Science Studies* 16, no. 2 (2003): 3–21; Nik Brown, "Shifting Tenses—From Regimes of Truth to Regimes of Hope?" *Configurations* 13, no. 3 (2007): 331–355; Mike Fortun, *Promising Genomics: Iceland and deCode Genetics in a World of Speculation* (Berkeley: University of California Press, 2008); Joan Fujimura, "Future Imaginaries: Genome Scientists as Socio-Cultural Entrepreneurs," in *Genetic Nature/Culture: Anthropology and*

Science beyond the Two Culture Divide, ed. Alan Goodman, Deborah Heath, and Susan Lindee (Berkeley: University of California Press, 2003); Kaushik Sunder Rajan, *Biocapital: The Constitution of Postgenomic Life* (Durham, NC: Duke University Press, 2006); and Charis Thompson, *Making Parents: The Ontological Choreography of Reproductive Technologies* (Cambridge, MA: MIT Press, 2005).

5 Catherine Waldby, "Stem Cells, Tissue Cultures and the Production of Biovalue," *Health: An Interdisciplinary Journal for the Social Study of Health, Illness and Medicine* 6, no. 3 (2002): 305–323.

6 Melinda Cooper has provocatively analyzed the imaginary of stem cell science in which life is rendered as its own surplus. Although her analysis is done in the context of embryonic stem cell research, the image of stem cell as surplus-generating life is still present in the field of stem cell research in general. See Melinda Cooper, *Life as Surplus: Biotechnology and Capitalism in the Neoliberal Era* (Seattle: University of Washington Press, 2008).

7 For these web pages, see "Diseases to which stem cells can be applied," http://www.bioinsurance.co.kr/cell/NowFuture.asp; and "The State of Art in Clinical Trials and Cases of (Successful) Treatments," https://www.kscb.co.kr/content/page.php?seq=23.

8 KSCB website, https://www.kscb.co.kr (emphasis in the original).

9 I am inspired by the concept of "surplus health" developed by Joseph Dumit as "the capacity to add medications to our life through lowering the level of risk required to be 'at risk' in the context of the growth of the pharmaceutical industry in the United States." In the case of bio-insurance, surplus emerges from the gap between the vitality of tissues in the present and its decreased vitality in the future, and becomes conceivable through the (presumed) capacity of stem cell technology to preserve the vitality that one will lose and with faith in the technological progress that would make stem cells useful in treatments in the future. For the notion of surplus health, see Joseph Dumit, *Drugs for Life: How Pharmaceutical Companies Define Our Health* (Durham, NC: Duke University Press, 2014).

10 I reconstructed a bio-insurance salesperson's one-on-one promotion scene here based on a note of the salesperson's lecture taken by a trainee that I could gather online. The address of the online post is not disclosed here because it was posted on a private blog, although it is publicly available.

11 Jeon Hongtaek, Park Myeongho, Yoon Seokmyeong, Song Yangmin, Jeong Hyeongseon, Bang Hanam, and Han Joohyeong, *Paeksesidae, ŏttŏke haengbokhage salgŏsinga? Wurinŭn kwayŏn paeksesidaerŭl maja ŏlmana chunbihago issŭlga?* [Centenarian era, how to live happily: How prepared are we for the coming centenarian era?] (Seoul: National Research Council for Economics, Humanities and Social Sciences, 2011).

12 The definition of *changnyŏnch'ŭng* is not clear. In one place, one of the authors used the word to refer to people in their fifties, and in another place, another author used it to refer to those in their forties and fifties. The only thing that is clear is that the authors use the term to emphasize their interest in the groups that are not-yet-old, the ones that require an immediate intervention to get prepared for the future.

13 Park Sang Chul, "100 nyŏn insaengŭl sŏlgyehara" [Plan the 100-year-long life], *Maeil Business Newspaper,* December 5, 2012, A39.

14 Choi Eunjin, Chae Heuiran, Park Cheongyeon, Jackie Green, and Katrin Engelhardt, *Ch'iryoesŏ yebangŭroŭi p'aerŏdaim chŏnhwan e ttarŭn kŏn'gang chŭngjin chŏngch'aek kaesŏn pang'an e kwanhan yŏn'gu* [Health promotion policy development in accordance with the paradigm shift from curative measures to preventive measures] (Seoul: Korea Institute for Health and Social Affairs, 2011).

15 The scholarly literature that concerns *mibyŏng* written in Korea began emerging around 2008 in a turn toward preventive practice in so-called Oriental medicine. Similarly, in China, medical anthropologists have observed the recent surge of *mibyŏng* (*weibing* in Chinese) as the antecedent of subhealth in the turn toward preventive practices in liberalized and commodified health care since 1990s. See Mei Zhan, "A Doctor of the Highest Caliber Treats an Illness Before It Happens," *Medical Anthropology* 28, no. 2 (2009): 166–188; and Mikkel Bunkenborg, "Subhealth: Questioning the Quality of Bodies in Contemporary China," *Medical Anthropology* 33, no. 2 (2014): 128–143.

16 "Yangbang, tongyang ŭihak, taech'e ŭihak hyŏpchinŭro chungjangnyŏnch'ŭng 'hoesaek chidae' ŏpsaenda" [Getting rid of the "gray zone" in middle-age (health) through collaboration among Western medicine, Oriental medicine, and alternative medicine], *HealthChosun,* June 22 2010, http://health.chosun.com/site/data/html_dir/2010/06/22/2010062201728.html.

PART III

PROSTHETIC ARTS

CHAPTER 8

Photography, Technology, and Realism in 1950s Korea

Hye-ri Oh

The dissemination of realism in Korean photography as it emerged in the 1950s was preceded by and coincided with a series of political, social, and cultural incidents, including liberation from Japanese colonial domination, the ideological conflict between South and North Korea, and the Korean War (1950–1953). Responding to major political and social upheavals, Korean photographers drew on their medium's capacity to record and claimed their commitment to social and historical circumstances. By resorting to the function of photographic technology, they tried to contribute to the rise of social consciousness. Korean photographer Yi Hyŏng-rok asserted that the essential capacity of photography is to record social reality and that photographers should never neglect historical events.[1] Similarly, photographer Pak P'il-ho articulated the mechanical reproduction of reality as the core function of photography.[2] Korean photographers began to conceive of photography explicitly as visual documentation, a mechanical process based on scientific developments in optics and chemistry. Yi emphasized that photography tends to convey moving and shocking images.[3] Documentary practices flourished as photographers sought to define the inherent quality of photography in their practice. The rise of photographic realism followed the demise of photographic pictorialism, which had been an international movement since the late nineteenth century and in which photographers responded against the claim that a photograph was nothing more than a record of reality and elevated the status of photography as an art. How pictorialist photography gained significance in Korea during the colonial era was more complicated than how it did in the West in that photographic practices were framed by Japanese cultural policies and the Korean response to colonial cultural hegemony. Pictorialist photography in Korea, with its emphasis on the subject matter of landscape and customs tinged with a lyrical atmosphere, can be seen as the manifestation of an ambivalent cultural statement of cultural negotiations and assimilation of the colonizer and the colonized during the colonial era.[4]

However, in formulating the shifting conception of photography, Korean realist photographers began to think of photography as a medium for actuality and tried to depict it in their work. Indeed, the dissemination of realist photography in Korea coincided with the global trend of realism during the post–World War II period. Simultaneously, the emergence of this conception of photography was a response to the changing political, social, and cultural circumstances of Korea in the 1950s. Following the thesis of the photographic historian John Tagg, who rejected a single totalizing history of photography, this chapter seeks to repudiate photographic realism as a "universal" and "a-historical conception."[5] At its outset, photography has been commonly accepted as the reflection of the real and the tangible.[6] Such an intrinsic and universal quality was never fixed, however, but instead has been continuously challenged by its technological changes and shifting political and social demands.[7] The divergent conceptions of photography and photographic technologies tend to be formulated in ways that respond to local cultural and political conditions and to intercultural dialogues in global contexts.

One direction in the study of photography is how the development of science and technology has had an impact on new ways of vision and determined social changes and cultural practices. However, moving beyond the reductionist view in technological determinism, this chapter explores the processes in which photographic technology gained its significance as an aesthetic medium and communication agency in society, and how Korean cultural and social contexts determined cultural values and meanings of realist photography. I argue that realism in Korean photography was a definite historical conception born out of a specific historical conjuncture. I conceive of photography not only as a cultural technology but also as a discursive construction. To this end, this chapter attends to the term *saenghwaljuŭi sajin* (life-ism photography) to refer to Korean realist photography. It highlights how the term maintained the importance of photography as a technical way of recording reality, as well as how it clustered around specific aesthetic, political, social, and cultural discourses. It illuminates the fact that Korean realist photography was deeply associated with prevailing contemporary thoughts and practices, including anticolonialism, anticommunism, antisocialism, and humanism. I contend that the ideas of realism and realist representation continued to be redefined concerning shifting discursive frames, the experience of the Korean War, the rise of an authoritarian political system, and cultural receptions and negotiations of the Western and Japanese documentary photography.

Realism, Photography, and *Saenghwal*

I start by examining the transformation of the concept of photography as much as possible with a focus on the conception of realism and its articulation in the term *saenghwal,* which played a critical role in determining subject matter and technique for a group of photographers. Realist photography demonstrated the aspiration of Korean photographers to deal with the harsh social realities of the post–Korean War era. A host of photographers including Limb Eung-Sik (Im Ŭng-sik), Yi Hyŏng-rok, Chŏng Pŏm-t'ae, Yi Kyŏng-mo, and Ch'oe Min-sik emphasized the mechanical ability of photography and produced shots of contemporary society. Koo Wang-sam, Limb Eung-Sik, and Yi Myŏng-dong contributed several articles and essays on the characteristics of photography and the mission of photographers to a variety of venues such as magazines, newspapers, and lectures. Koo Wang-sam advocated realism and articulated realist photography in the magazine *Sajin Munhwa* in 1956 as follows:

> Realist photography is uncontrived and "absolute" snapshot photography. It captures the "decisive moment" (*kyŏlchŏngjŏk sun'gan*), as Henri Cartier-Bresson called it. . . . Understanding the significance of incomprehensible life is a prerequisite for expressing reality. Photography portrays reality and embodies the spirit of the times. Realist photography is distinct from previous photographic practices. It demands a new innovative vision and the refinement of technological skills.[8]

No doubt Koo's statement demonstrated the changing notion of photography that most contemporary photographers in Korea shared. He characterized realist photography as a perfect snapshot without manipulation, conscious of the camera as an instrument of "objective" documentary. He championed the direct relationship with the camera and the object to produce realist photographs. Since the invention of photography in 1839, the camera continued to be considered an objective picture-making machine characterized by its indexical quality, that is, a one-to-one correspondence between the object and its representation. Koo, emphasizing "the spirit of the times," underlined that realist photography should keep abreast of the historical confluence of the ideas and practices of the moment. Indeed, such recognition on the importance of historical contexts became the object of study among photography theoreticians. For instance, John Walker stresses context as a crucial factor of a particular photographic meaning. A photographic image is often open to multiple interpretations. According to Walker, humans as social beings are inevitably affected by social conditions, and therefore a photographic image is particularly potent and its interpretation depends on the contextual circumstances.[9] By interrogating the referential quality of photography, John Tagg challenges the notion of photography as such, asserting the multiplicities of photography by arguing that the medium has no fixed function and that therefore heterogeneous institutional spaces in different historical circumstances determine its multiple functions.[10] Joel Snyder repudiates a natural relation of resemblance between objects and representational images based on an ontological approach and instead argues for picturing vision in realistic representation as the product of social and cultural habituation, judgment, and inculcation throughout history.[11] In his remarks, Koo outlined the understanding of specific historical context as a prerequisite for successful realist photography, and hence diverse functions and values of photography to deal with the changing social circumstances.

Further, Koo indicated that the emergence of photographic realism was a response to previous photographic practice, a convention that had implied the photographic pictorialism that became widely circulated in Korea as *yesul sajin* during the first half of the twentieth century. Photographic pictorialism refers to a part of the international photographic movement to confer photography with the status of art from the late nineteenth century to the first decades of the twentieth. Since its inception, photography had mainly been identified as business, science, and technology rather than as an art form. Photography had to prove that it was not mere mechanical registration of a moment, but instead an artistic medium that represents individual expression, creativity, and authorship. To this end, photographers strove to create an artistic mood in their works by integrating atmospheric effects, tonality, composition, and the beauty of the subject matter into photographs. First introduced into Korea in the late nineteenth century, photography was conceived as a product of Western science and technology and hence was appraised as a way to strengthen and civilize the country in conjunction with the burgeoning discourse of "civilization" and "enlightenment."[12] Following the international currents, Korean photographers participated in breaking with its obligation as a utilitarian mechanical medium and witnessed the prevalence of pictorialism in modern Korea. However, the dissemination of Korean pictorial photography to a great extent came in conjunction with Japanese colonialism and unfolded differently than it did in the West. After the March First independence movement in 1919, the Government-General of Korea needed to heighten the efficiency and sophistication of the colonial governance to deal with the nationalist fervor of the Korean masses against repressive military rule. Japanese colonial authorities shifted their focus from military to culture. During cultural rule (*bunka seiji*) from 1919 to 1931 in colonial Korea, photography was promoted as art through several venues, including mass media, exhibitions, and photography contests. The contemporaneous notion of "culture" emphasized the significance of spirit, self-formation, and individualism. Korean pictorialist photography, translated into *yesul sajin* in Korean, was tinged with personal and individualistic contemplation. It also determined the characteristics of Korean pictorialist photography with a focus on the theme of nature, indigenous scenic spots, or customs, which was preferable as a guideline in the photographic exhibitions and contests during the colonial era. Indeed, the discourse of culture converged with the currency of art (*yesul*) at various points within the colonial circumstances. The Government-General of Korea disseminated the currency of the terms "art" and "culture" in response to the weight placed on the pursuit of individual spirituality and self-expression, which did not posit threatening activities and thoughts to colonial rule.[13]

The significant figures of realistic photography in the 1950s began their photographic career in the colonial era and therefore immersed themselves in pictorialist practices.[14] One critical social and cultural endeavor in Korea after liberation was to overcome colonial cultural legacies. Given their heightened social consciousness, photographers tended to think of realism as an alternative to old-fashioned stylistic photography, disparagingly known as salon photography in Korea.[15] Koo Wang-sam criticized salon-style photography, insisting that

Fig. 1 Limb Eung-Sik, *Retuning Home*, 1932. National Museum of Modern and Contemporary Art, Korea.

> salon photography was not the result of creativity but was just a technical play through shallow skills. The authentic photographic art demonstrates the critical ideas of photographers, the ideas of the times, and the historical concerns but salon photography never dealt with them.[16]

Realism permeated in the fundamental value of photography as a part of cultural tendencies to counter the colonial cultural legacies. For example, the ninth exhibition of the Korean Photographers' Association (Sajin chakka hyŏphoe) in 1957 displayed a large number of realist works in which the human being became the major motif. The photographers demonstrated ordinary people in their everyday lives.[17] They achieved realism by abandoning their familiarized obsolete way of picture making for personal, individual, and lyrical expression prevalent in the colonial era. In the same vein, Koo stressed the social function of photography insistently, claiming that photographers should contribute to the construction of Korean national culture and keep abreast of the demands of the society. Photography was supposed to promote the public interests and photographers showed their commitment to accommodate social demands.[18]

Despite the effort to overthrow the impacts of Japanese culture, what is evident in Koo's articulation is that the notion of Korean realist photography was analogous to that of Japan. The axiom for "absolutely unstaged" and "snapshot photography" was pronounced in the thesis of Japanese photographer Domon Ken (1909–1990), a leading figure in advocating socialist realism in postwar Japanese photography.[19] Several Korean photographers acknowledged their affinity with the currents of Japanese realist photography even after liberation.[20] In the aftermath of World War II, the need for critical attention to social situations was urgent. To respond to social needs, Domon criticized photographic pictorialism for having no positive function in society.[21] Domon argued that "photographic realism is the only true photography for contemporary times, and that one photograph is its substantiation."[22] He believed that only photographic realism has the potential to connect directly with social reality, and the mechanism of the camera is the ideal agent for its aim. Domon expressed his faith in the technological power of photography to capture the moment with no artificial intervention when he pressed the shutter. Domon's thoughts on photographic realism that demanded the fidelity of photography to social phenomena provided a clear opposition to pictorial photography and suggested a different path for photography to follow.[23]

It was Limb Eung-Sik who coined the term *saenghwaljuŭi sajin* to refer to the conceptual category of Korean realist photography in the 1950s.[24] *Saenghwaljuŭi sajin* proposed to envision the postwar condition and human life in a humanistic approach as well as to search for the social "truths."[25] Limb encouraged photographers to look squarely at life and to represent devastating social conditions and daily life as subjects of "human interest" in the aftermath of the Korean War in a truthful way, without intervention or exaggeration.[26] For example, Limb's well-known *Looking for a Job* is a black and white photograph of an unemployed man. However, the subject's grandiloquent posture seems to show a discrepancy from Limb's original aim of featuring ongoing social severity in that such a framing prevented it from revealing the predicament of the times.

A close-up image of an unemployed man seems to fail in conveying visual information of the situation the figure is undergoing. It shows no more than sheer youth partially hidden under a cap and a leaning figure against a wall along the street. It is not hyperbole that such an image focusing on form, style, and composition evidenced the long-held pictorialist style of the colonial era that Limb followed so closely. Limb himself

Fig. 2 Limb Eung-Sik, *Looking for a Job*, 1953. Gelatin silver print. National Museum of Modern and Contemporary Art, Korea.

confessed that it was not easy to abandon the habit of pictorialist photography and its focus on formalistic elements such as style, form, or light.[27] Although the photograph seems to fail to convey a graphic image of political and social disturbance of the period, it functions as a strong indictment of Korean society. Undeniably, unemployment became a critical social issue after the Korean War. The portrait of an unemployed man and the sign "looking for a job" hung in front of his chest echoes unmistakably the socially immediate concern. Limb continued to confront the events of the 1950s and demonstrate the experience of Korean people. In *War Orphan,* Limb turned his camera lens to reality in an empathic and straightforward way to depict the crucial issue of thousands of war orphans. The Korean government failed to face these social problems promptly. Because of this, Korea ended up being stigmatized as the country for overseas adoptions. Limb's choice of subject matter could be seen as timely and unmitigated enough to attract attention from the population at large.

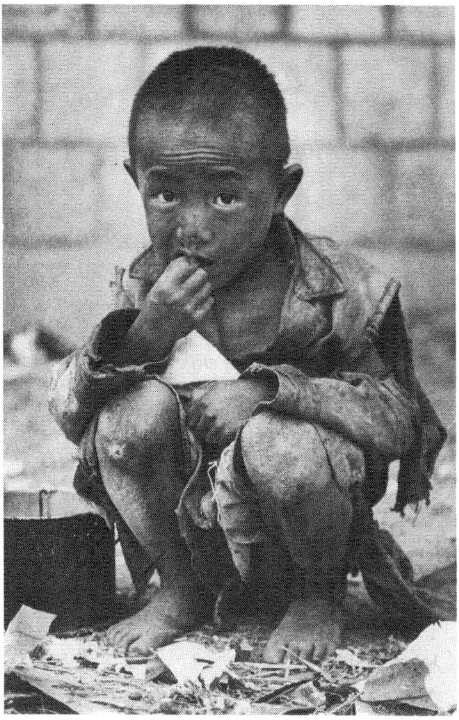

Fig. 3 Limb Eung-Sik, *War Orphan,* 1950. Gelatin silver print. National Museum of Modern and Contemporary Art, Korea.

Limb's thought on realism and his interest in social issues was formulated from his experience as a photojournalist during the Korean War. He suggested that photographers capture humanistic narratives as a crucial component of *saenghwaljuŭi* realist photography, which is comparable to the journalism approach.[28] In a review on the eighth exhibition of the Korean Photographers' Association in 1956, Limb declared the sweeping victory of *saenghwaljuŭi sajin* over lyrical photographic conventions:

> In the past, photography did not adequately represent the life and the emotion of the general population [*minjung*]. It displayed no insights into reality. What photography did was to portray the natural beauty with an emphasis on formal elements. However, the experience of the Korean War forced the younger generation to turn their eyes to the harsh conditions of life. It achieved the objective and realist expression of everyday life.[29]

The conception of objective, realistic, and socially engaged *saenghwaljuŭi sajin* was privileged as an alternative to the hitherto prevailing pictorial, romanticized, lyrical, and individualistic salon-style photography to deal with serious political, social, and cultural situations. What meaning was implicit in the choice of the term *saenghwal* in designating Korean realist photography? Limb acknowledged that the government was hostile to the term realism because of its association with socialism and radicalism.[30] As art historian Ch'oe Yŏl remarked, Korean art practices and its criticism of the 1950s were circumscribed, to no small extent, by two dominant imperatives: antisocialism and anticommunism and high regard for Western capitalism and its culture.[31] Supposedly, the gentle and nonconfrontational portrayal of domestic problems became a bulwark against the Korean authoritarian government that was extremely antagonistic to socialist perspectives. The photographers tried to avoid public controversy and surveillance.

The term *saenghwal* was widely used during the colonial era and gained importance in conjunction with socialist and materialist thought.[32] Socialist thoughts began to be disseminated by the left wing during the colonial era. As early as the 1920s, realism began to draw attention from a group of artists who emphasized the social function of art in resolving the grim and bitter conditions of the general population (*minjung*) in colonial Korea. In particular, the socialist group PASKYULA—founded in 1923 by Kim Pok-jin, Pak Yŏng-hŭi, Yi Sang-hwa, and Kim Ki-jin—tried to deal with the shifting social and political circumstances. It stressed the significance of art for the *minjung,* who were an integral social force after the March First independence movement in 1919.[33] Art for the *minjung* needed to portray the life of Koreans under severe colonialist oppression. As a politically and socially charged concept, realism became significant among artists and intellectuals.[34]

In an attempt to overcome what they saw as a decadent individualistic culture, socialists attended to everyday life conditions of ordinary people and found an alternative in the term *saenghwal* to allude to realism. They launched such magazines as *Kaebyŏk, Sin saenghwal,* and *Chosŏn chigwang* and disseminated the usage of *saenghwal.* Kim Ki-jin and Pak Yŏng-hŭi claimed art for *saenghwal,* not art for art's sake. To them, whereas art for art's sake is related to individualistic, decadent, and playful visual representation, art for life is constructive, productive, collective, and social.[35] Kim Ki-jin argued that no art is possible without *saenghwal.* In his thought, art should demonstrate the deprived and tragic life of people.[36] The poet Yi Sang-hwa argued that the value of literature and art was determined by how concretely it represented *saenghwal.*[37] One of the crucial mottos of the Korea Artista Proletaria Federation was "*saenghwal* is art and art is *saenghwal*."[38] Yŏm Sang-sŏp argued that "art is born out of *saenghwal*. There is no lively and valuable art without conveying *saenghwal* by infiltrating life in depth and capturing reality."[39] Western-style painter Kim Chŏng-t'ae asserted that "art aims to improve *saenghwal*. Art is no mere hobby or the academic subject but is deeply concerned with *saenghwal*."[40]

Moreover, art should take the form available to anyone and can be understandable to any members of society. Presumably, photographers internalized the prevalent cultural rhetoric and applied to spell out the characteristics of their representational practices. The naming of postwar realist photography was subject to the reconfiguration of the concept of *saenghwal* to accommodate the demands of the politically, socially, and culturally specific conditions of the 1950s after the war. Korean photographers emphasized the notion of *saenghwal* and incorporated it into their conceptualization of photography as a cultural agency to implement the function of the "witness of the actuality." Even though realism in the visual arts has generally been regarded as a movement grown out of Western visual culture, we can challenge the generalized assumptions about realism and its related practices established in the Western tradition. In Korea, the consciousness of reality emerged as a part of socialist ideas during the colonial era in order to overthrow the painful colonial reality. The notion of *saenghwal* continued its currency in general social and cultural discourse in the aftermath of the Korean War. *Saenghwal* was chosen as an alternative term to realism in the 1950s and conveyed the social function of art to communicate reality.[41]

After liberation from Japanese occupation and the establishment of the United States Army Military Government in Korea (USAMGIK) in 1945, Korea became a more ideologically split society, which led to national division and a devastating civil war. In 1947, USAMGIK and the United Nations endorsed the foundation of the Republic of Korea (ROK). In response, North Korea established the Democratic People's Republic of Korea (DPRK) after holding elections. The ideological and political antagonism between the ROK and DPRK resulted in the Korean War on June 25, 1950. A truce negotiation was signed on July 27, 1953. The control of the USAMGIK over South Korea was pervasive in repressing leftist ideology and provided long-term restraint in the visual culture of South Korea as well as in political and economic policies. Although President Syngman Rhee of South Korea promoted a campaign of anticommunism and antisocialism, the government was susceptible to social critique and did not permit left-wing ideology.[42] The government loathed the term "realism" on the grounds that realism was associated with socialist theory in its origins and was thus incompatible with the government's ideology.[43]

In these circumstances, the choice of *saenghwal* seems to account for photographers' intentional response to the atmosphere prevalent in politics and society. Authoritarian government policies formalized anticommunism and antisocialism, and cultural practices became a target of surveillance.[44] Asserting consistent interest in representing social realities, Limb Eung-Sik skirted realism in photography by adopting the term *saenghwaljuŭi*.[45] The euphemism helped photographers avoid surveillance of a government sensitive to anticommunism and antisocialism.

Such self-restraint was also reflected in Koo Wang-sam's incongruous remarks. Koo stressed that realist photography should faithfully represent the actual. Meanwhile, he contended that photography should not be associated with ideology and should not function as a tool for socialist realism.[46] The features of Korean realist photography, such as the choice of subject matter, styles, or frame, can be seen as a manifestation of deeply rooted tensions of ideological conflict.

Soon after liberation, a variety of left-wing groups were founded. One, Chosŏn sajin tongmaeng (Alliance of Chosŏn Photography), was an affiliate of Chosŏn munhwa tanch'e ch'ong yŏnmaeng (Federation of Chosŏn Cultural Groups). With twenty members, including Yi T'ae-ung and Kim Chin-su, the group sought to contribute to national culture.[47] Koo Wang-sam served as head of the executive committee in the regional group in Taegu.[48] Photographer Lim Seok Je (Im Sŏk-che) was also a member and held the first photography exhibition at what was then Tonghwa department store in August 1948.[49] The show was technically a single artist exhibition but also displayed the photographs of Chosŏn sajin tongmaeng members such as Yŏm Pyŏng-t'aek, Yi T'ae-ung, Kim Chin-su, and a nonmember realist photographer Im In-sik.[50] Lim Seok Je's primary theme was food imports under the regime of the USAMGIK and portrayed the deprived and workers in his works.

It is impossible to deny that Korean photographers demonstrated their commitment to reflecting society and accommodating social progress. However, the aspiration for the realism of the photographers in the 1950s was, to a certain degree, ambiguous, and broad and had no rigid ideological complexion. It remained very much at the emotional level and did not reveal analytic insight. Sometimes the photographers did not follow the distinction between the two ideological systems. Most of the

Fig. 4 Lim Seok Je, *Imported Foods*, 1948. Gelatin silver print. Private collection.

leftist Chosŏn sajin tongmaeng members—Hyŏn Il-yŏng, Yi T'ae-ung, Lim Seok Je, and Kim Chin-su—were also involved with the right-wing group, Chosŏn sajin yesul yŏn'guhoe.[51] Compared with the activities of the radical left-wing groups in such spheres as literature, art, and film, the leftist activities of photography were somewhat equivocal. Historian of photography Lee Kyŏng-min assessed that this was due to the lack of intelligentsia working in the field of photography.[52] Besides, very plausibly, the institutional apparatuses remained at the embryonic stage and did not guide photographic practices. Institutions such as Seoul National University, Ewha Womans University, and Pusan University had just begun to offer photography courses in the 1950s.[53] To carry a Leica or Contax camera became a trend in the 1950s.[54] Several photography clubs were active, but the activities remained at the amateur level. Photographic practices were thus not built on any theoretical foundation or clear ideological consensus.

The Korean War, Social Consciousness, and Photographic Technology

Photography was considered an effective visual medium for everyday life in that it tends to verify and substantiate reality. The photographers had faith in the objective power of a photographic machine and fidelity to an actual event in its final product. As Lim Seok Je argued,

> Realism is related to the event that happens. Consequently, it satisfies the demands of journalism and photographic art. The camera turns its lens to the subject matter in order to understand humanity. To rediscover humanity in society is what contemporary photographers strive to achieve. The camera lens portrays the external appearance and living environment, which enables us to infiltrate into the inner aspects of human beings. It is the principal function of realism that can be achieved by camera tools.[55]

The technical process of photography that draws on its chemical replication of actuality has long been held to be an inherent feature. However, photographic realism is not the achievement of a mere optical device. Instead, it is fulfilled by the person who chooses the subject matter and the moment to click the shutter. A photographer's subjectivity contributes to the production of the work. Not neglecting the fact that photography is intrinsically a mechanical art, Koo Wang-sam claimed that photographers integrate their personal experience, individual positions, and social consciousness into their production.[56] Similarly, responding to the claim that photography is not art because its expression is the product of mechanical device, Limb Eung-Sik contended that such an argument resulted from an incomplete understanding of photography. Photographic expression, he noted, emanates through a photographer's subjective sensibilities. Therefore, the choice of subject, light, distance, angle, and aperture completes creative photographic practices.[57] To him, the aesthetic expressions of photography originate in the choice of the subject as a social being, not from the mechanical tools needed to produce the photograph.[58] In the same vein, Yi Myŏng-dong asserted that "unlike other machines, the camera does not perform itself repeatedly. Instead, it operates through the action and consciousness of who controls it. It allows photographers to achieve artistic creativity in their practices."[59] His statement seems to repudiate the presupposition that photographic representation is closely tied to a particular technology. Instead, as he made clear, reality can be depicted in terms of the connection between a subjective position and an extensive discursive system of representation. No doubt the personal experience and subjective position nurtured in the historical and cultural circumstances are an indispensable component of photography as a communication agency and aesthetic expression.

Most realist photographers found their inspiration from their direct involvement in the Korean War and a series of ensuing social incidents. Yi Hyŏng-rok served as a documentary photographer in the war and recorded the casualties of the combat of Naktong River. Limb Eung-Sik covered the war as a photojournalist and had opportunities to meet Hans Walker, Margaret Burke-White, and Carl Mydans, each of whom worked for the magazine *Life*.[60]

Han Yŏng-su remarked that the experience of a traumatic war had a lifelong impact on his photographic practice.[61] As a photojournalist, Yi Kyŏng-mo also covered the Korean War and the 1948 Yŏsu-Sunchŏn rebellion for the newspaper *Homan sinmun* (now *Kwangju ilbo*). His professional career enabled him to know precisely the mission to convey the news with the telling photographs to the expectant readers. Yi also worked

Fig. 5 Limb Eung-Sik, *Korean War, Seoul*, 1950. Gelatin silver print. National Museum of Modern and Contemporary Art, Korea.

for *Kukche podo* (Pictorial Korea),[62] the first Korean pictorial journal, launched in November 1945 by editor Song Chŏng-hun, who recognized the communicative power of photography and social function of the visual images. In the first issue, Song made it clear that the mission of the journal was to contribute to the nation and society by delivering Korean national culture and the current news. The journal was published in Korean and Chinese characters, and in English, which aimed at an international cultural exchange, along with constructing Korean national culture.[63] It carried the works of many realist photographers including Limb Eung-Sik, Hyŏn Il-yŏng, Kim Chŏng-rae, Lim Seok Je, Chŏng Pŏm-t'ae, Yi Hae-sŏn, and Ch'oe Kye-bok. Notably, Kim Han-yŏng and Yi Kyŏng-mo worked for *Kukche podo* under exclusive contract for thirteen years and five years, respectively.[64] The photographers understood the photographs as the visual testament of their extant social concerns. On the basis of its visual credibility, they considered photography tangible proof of ongoing social issues and an agent for communication to facilitate the viewer's attention to social reality.

Fig. 6 Yi Hyŏng-rok, *Shoe Seller, Namdaemun Market*, 1957. Gelatin silver print. National Museum of Modern and Contemporary Art, Korea.

Humanism and the Reception of *The Family of Man*

The photographers' group Sinsŏnhoe is worthwhile to remember given that the group's idea and practices epitomized the original vision of *saenghwaljuŭi sajin* despite the group's short-lived activities.[65] Founded in 1956, it was a group of nineteen amateur realist photographers, among them Yi Hyŏng-rok, Cho Kyu, Yi An-Sun, Son Kyu-mun, Chŏng Pŏm-t'ae, Yi Hae-mun, Han Yŏng-su, and An Chong-ch'il.[66] The word *sinsŏn* signifies a "new line" or "new stream"—*sin* meaning "new" and *sŏn* meaning "line" or "stream."[67] The underlying objective was to follow the international stream of realism and to pursue an unconventional approach, diverging from pictorial photography. The members often commented on the Photo-Secession group led by the American photographer Alfred Stieglitz (1864–1946) as the inspiration for their photographic practices.[68] Following the international trend of modernism in visual arts, the Photo-Secession group searched for the medium's intrinsic aesthetic quality of photography by portraying form and tonal gradation in the real world.[69] Likewise, the members of Sinsŏnhoe underscored what they called "decisive moments" of everyday life in straight black and white photography. The first group exhibition was at the gallery of the Tonghwa department store in April 1957 and presented in two sections. The first displayed works on a variety of topics, and the second displayed lively market scenes taken at different times—dawn, midday, and night (*sijang-ŭi saengt'ae*). The show offered a fresh photographers' outlook on reality. According to an exhibition review in the magazine *Sajin munhwa* (Photography culture), the group strove to portray contemporary society and challenged long-held stylistic conventions. Rather than formal elements such as composition, light, or angle, the photographs showed the scenes and humanity of the lively markets vigorously.[70] The photographers conceded realism as the direction of photographic art. Limb Eung-Sik, whose ideas provided the foundation of Sinsŏnhoe, appraised that the photographs of the group's members aimed at capturing human life in a single moment.[71] However, despite their efforts, the members of Sinsŏnhoe confessed that it was sometimes hard to move beyond the force of their habit, and their work was bound to reveal vestiges of the pictorialist style. Yi Hyŏng-rok pointed out that Korean realist photography in the 1950s stood midway between salon-style and realistic photography.[72]

The photographer Yi Hae-sŏn stressed humanism in photography and encouraged other photographers to deal directly with human life. She also recommended that they should nurture their personality along with the techniques, skills, and creativity to deliver the essence of humanity.[73] Photographers tended to record the hard conditions of life with an emotional and sympathetic approach, giving Korean photography quality of emotional realism. The interest in the humanistic perspective was also confirmed in the success of the 1957 exhibition *The Family of Man* held at the Kyŏngbok Palace Gallery from April 3 to 28.[74] Most realist photographers in Korea cited the exhibition as the valuable source of their inspiration in photographic practices of the times.[75]

The 1955 show *Family of Man* was mounted by Edward Steichen (1879–1973) in the Museum of Modern Art in New York and ran from January 24 to May 8. Commissioned by the US Information Agency (USIA), the exhibition then circulated worldwide, including England, Germany, Belgium, Italy, France, Japan, India, Iran, Burma, Indonesia, Australia, Venezuela, Chile, and Korea.[76]

The most successful and influential exhibition in the history of photography, the *Family of Man* attracted 1,200,000 spectators from 1955 to 1959 in the United States and Canada and 7,500,000 in other countries.[77] A total of 2,777,000 volumes of the *Family of Man* catalog were sold

Fig. 7 The installation view of the *The Family of Man* exhibition at the Museum of Modern Art, New York, 1955. Museum of Modern Art Archives. Photo by Rolf Petersen.

between 1956 to 1979.[78] According to USIA records, the show attracted 420,000 people in Korea.[79] Steichen received two million photograph submissions from across the world and scrutinized already published magazines such as *Life* and *Look* as well as the photojournalist archives of *Magnum*.[80] Ultimately, he screened 503 photographs by 273 photographers from sixty-eight countries. The selected photographs were divested of their original contexts and relocated to visualize the essential and universal elements of human life including birth, love, wedding, worship, joy, faith, happiness, and hopes.[81] As Carl Sandburg wrote in the prologue of the *Family of Man* catalog, the show aimed at demonstrating that "there is only one man in the world and his name is All Men. There is only one woman in the world and her name is All Women. There is only one child in the world and the child's name is All Children."[82] Drawing on the capacity of camera testimonial, the core narrative of the exhibition was to represent dramatic scenes of the "grand canyon of humanity"[83] and to focus on the commonality of life around the world. The theme was compelling enough to appeal to the viewers' emotions and sentiment, and thus was received well by the public. The exhibition proved that photography can depict the shared experience and is a potent and valuable tool of mass communication.[84]

The exhibition was the product of specific social or political considerations. The early 1950s was a period of reconstruction from the devastation of World War II. The United States began to gain political and cultural leadership in the changed international order because it was relatively unsullied by the two wars. During the Cold War, art functioned as a privileged instrument that could convey the consciousness and sentiment of people. The concept of the family was expected to reinforce global kinship, and the *Family of Man* suggested an idealistic unification of the world as a single unit of the family. The exhibition not only enabled people to forget about the disheartening past but also to believe in the fundamental righteousness of humanity, giving them hope and offering promise. However, the exhibition became an object of controversy among critics, especially in light of its ideological ambition. It tried to bind the world as a family unit by divesting the photographs from a variety of sources of the original messages.[85] As photo-critic Allan Sekula remarked, photography functions as a way of looking at reality straightforwardly based on the power of the scientific aspect of photography. However, Sekula continued to emphasize that "the worldliness of photography is the outcome, not on any immanent universality of meaning, but a project of global domination."[86] The international tour of the *Family of Man* was viable under the auspices of USIA, which had been founded in 1953 to promote communication with foreign countries and developed its foreign policies and international relations through communication media. The exhibition also became a competent medium in the Cold War fight for cultural hegemony.[87]

Regardless of the original ideological complexion, the *Family of Man* acquired new significance in each local condition. As John Walker asserts, the repositioning of visual representation in a new contextual background does not necessarily generate a different meaning of a photograph. However, a context change resulted in the change of focus and interpretation depending on the receiver's urgencies.[88] How did Koreans understand and negotiate the cultural experience of the *Family of Man* show in their specific historical conditions? What parallel social and cultural discourses enabled the exhibition to succeed in Korea as well as in the United States? Before the exhibit opened in Korea, several mass media introduced it, along with a biography of Edward Steichen, to Korean readers.[89] The exhibition in Korea showed 501 photographs.[90] Photography critic Yi Myŏng-dong remarked that "art is the expression of human life and photography portrays the truthful aspects of life such as joy, sadness, love, hatred, and hunger."[91] Yi gave prominence to humanistic value as a crucial component of visual art and photography. Art critic Kim Yŏng-ju also remarked that humanism expressed human existence and conditions.[92] Humanism became a dominant cultural discourse in 1950s Korea. As humanism and existentialism provided the necessary philosophical guidance in recovering humans from the traumatic experience of World War II, so did humanism serve as a critical solution to deal with the devastating experience of the Korean War.[93] To draw attention to human life and its condition became a critical task of visual representation. By connecting humanism with realism, Yun Hŭi-sun claimed that art should convey hope, faith, and joy.[94] Realist photographic practices corroborate such a connection in an attempt to look straightforwardly at reality. Chŏng Pŏm-t'ae, who participated in the group Sinsŏnhoe, argued this way:

> Photography can record permanently every incident, including natural disasters and human-made disasters. . . . Every incident about humans is worthwhile to be recorded. . . . Through such images, photography can elicit social consensus and sympathy. Photography, as a category of art, should convey hopes and happiness to human beings.[95]

Korean photographers interpreted that faith in humanity was substantiated by the documentary images in the *Family of Man* exhibition.[96] Limb Eung-Sik asserted that the exhibition clearly demonstrated the direction of Korean photography in that it showed a prototype for Korean photographers' interest in humanism.[97] Yi Kyŏng-mo remarked that documentary photographic practices gain popularity after a show.[98] Korean photographers believed that the exhibition attested to photography as a tool for a visual record. In the end, the convergence of the parallel discourses of the West and Korea on humanism drove the success of the exhibition in Korea.

The weight of realism in photography became an international currency of the mid-twentieth century. Focusing on the meaning of realist photography in 1950s Korea, this chapter attempted to explicate the conditions and the particular moment in which photographic realism gained significance. The power of photography as technological reproduction of actual moments was widely acknowledged. The chapter illuminated the convergence and the divergence of photographic practices and parallel discourses of visual representation in a specific social context—the transnational cultural relationship between Korea and the West and between Korea and Japan during the postcolonial era. It showed the changing political, social, and cultural discursive frames after the Korean War, the struggle against colonial cultural legacies, and how different ideological systems intervened in the construction of the conception of photography. The interaction between Korea and Japan, even after the colonial era, and the newly established political and cultural relationship between Korea and the United States, complicated the concept of realism in photographic practices in Korea in the 1950s. Korean photographers portrayed contemporary life conditions sympathetically, whose approach became characteristic of Korean photography. This formulation developed within a complex interplay of colonial cultural residue, an authoritarian political regime, the assimilation of Western photographic practices, and the social and cultural needs of Koreans. Moving beyond the history of photography as technological determinism, this chapter highlights photography as "practice," which works on specific materials within a specific social and historical context and for a specific purpose.[99] The chapter challenged the Euro-American–centric historical perspectives on photography and provides an opportunity to link the changing function of photographic technology to historical conditions in different and complicated ways.

Notes

1 Im Yŏng-gyun, *Sajin'ga wa ŭi taehwa: Han'guk sajin ŭl kaech'ŏk han wŏllo sajin'ga 8in kwaŭi taedam* 3 (Seoul: Nunbit, 1998), 82.
2 Pak P'il-ho, *Sajin ŭl mal handa* (Seoul: Sigak, 2003), 24.
3 Im, *Sajin'ga wa ŭi taehwa,* 82.
4 On the rise of art photography, see Hye-ri Oh, "The Concept of Photography in Korea: The Genealogy of Korean Conception of the Korean Conception of *Sajin* from the late Chosŏn Dynastic Period through Japanese Colonialism" (PhD diss., Binghamton University, State University of New York, 2014), 133–209.
5 John Tagg, *The Burden of Representation* (Minneapolis: University of Minnesota Press, 1993), 175.
6 John Tagg, "The World of Photography or Photography of the World," in *The Camerawork Essays: Context and Meaning in Photography,* ed. Jessica Evans (London: Rivers Oram Press, 1997), 67.
7 James Elkins, *Photography Theory* (New York: Routledge, 2007), 4–5.
8 Koo Wang-sam, "Sadan yugam ilje," *Sajin munhwa,* June 1956, 75. Koo Wang-sam used the term *kyŏlchŏn jŏk sun'gan.* It was a direct translation of the term "decisive moment" coined by the French photographer Henri Cartier-Bresson to emphasize photography's power of capturing a given moment. The work of Cartier-Bresson was well known in the 1950s and inspired Korean realist photographers.
9 John A. Walker, "Context as Determinant of Photographic Meaning," in Evans, *The Camerawork Essays,* 52–57.
10 Tagg, *Burden of Representation*, 63.
11 Joel Snyder, "Picturing Vision," *Critical Inquiry* 6 (Spring 1980): 499–526.
12 Hye-ri Oh, "Translating 'Photography': The Migration of the Concept of *Sajin* from Portraiture to Photography," *History of Photography* 39, no. 4 (November 2015): 383.
13 Oh, "Concept of Photography in Korea," 133–209.
14 Im, *Sajin'ga wa ŭi taehwa*, 88.
15 Yi Myŏng-dong, "Riŏllijŭm ŭi t'ongil- sajin chakka hyŏphoe kuhoejŏn ŭl pogo," *Tonga ilbo*, November 16, 1957; and Pak, *Sajin ŭl mal handa*, 274.
16 Koo, "Sadan yugam ilje," 75.
17 Yi Myŏng-dong, "Riŏllijŭm ŭi t'ongil- sajin chakka hyŏphoe kuhoejŏn ŭl pogo," *Tonga illbo*, November 16, 1957.
18 Koo, "Sadan yugam ilje," 74–75.
19 Ivan Vartanian, Akihiro Hatanaka, and Yutaka Kambayashi, ed., *Setting Sun: Writings by Japanese Photographers* (New York: Aperture, 1992), 22–23.
20 Im, *Sajin'ga wa ŭi taehwa*, 23, 103.

21 John Szarkowski and Shoji Yamagishi, ed., *New Japanese Photography* (New York: Museum of Modern Art, 1974), 9–10.
22 Ibid., 27.
23 Iizawa Kōtarō, "The Evolution of Postwar Photography," in *The History of Japanese Photography*, ed. Anne Wilkes Tucker et al. (New Haven, CT: Yale University Press, 2003), 210–214.
24 Limb Eung-Sik [Im Ŭng-sik], "Sadan ŭi hyŏnjae wa changnae- saenghwal chuŭi sajin ŭi saengsan ŭl wihayŏ," *Kyŏnghyang sinmun*, June 9–10, 1955; and "Saenghwal chuŭi sajin ŭi sŭngni—Pyŏngsinnyŏn sadan hoego," *Kyŏnghyang sinmun*, December 19–20, 1956. Both articles are included in Limb Eung-Sik, *Limb Eung-Sik (Im Ŭng-sik) hoegorok: Naega kŏrŏ on Han'guk sadan* (Seoul: Nunbit, 1997), 243–246, 279–282.
25 Limb Eung-Sik, "Han'guk sajin'gye ŭi hyŏnhwang- chakhwa kyŏngyang ŭl chungsim ŭro," *Sŏul sinmun*, August 1959; and Limb, *Im Ŭng-sik hoegorok*, 287.
26 Limb Eung-Sik, "Sahyŏpchŏn ŭl pogo—saenghwal chiksi," *Chugan Hŭimang*, December 1957; and Limb, *Im Ŭng-sik hoegorok*, 277.
27 Im, *Sajin'ga wa ŭi taehwa*, 72.
28 Limb Eung-Sik, "Sadan ŭi hyŏnjae wa changnae-saenghwal chuŭi sajin ŭi saengsan ŭl wihayŏ," *Kyŏnghyang sinmun*, June 9–10, 1955; Limb, *Im Ŭng-sik hoegorok*, 246.
29 Limb Eung-Sik, "Saenghwal chuŭi sajin ŭi sŭngni—pyŏngsinnyŏn sadan hoego," *Kyŏnghyang sinmun*, December 19–20, 1956; and Limb, *Im Ŭng-sik hoegorok*, 279–280.
30 Im, *Sajin'ga wa ŭi taehwa*, 103.
31 Ch'oe Yŏl, *Han'guk hyŏndae misul pip'yŏngsa* (Kyŏnggido, P'ajusi: Ch'ŏngnyŏnsa, 2012), 154.
32 Yi Ch'ŏl-Ho, "Sin Kyŏnghyangp'a pip'yŏng ŭi nangman chuŭijŏk kiwŏn," *Minjok munhaksa yŏn'gu* 38 (2008): 235–237.
33 Ch'oe Yŏl, *Han'guk kŭndae misurŭi yŏksa* (Seoul: Yŏrhwadang, 2006), 166; and Chŏng Chu-yŏng, "P'asŭk'yura ŭi misullon," *Misul iron'gwa hyŏnjang* 5 (December 2007): 43–80.
34 On the reception of realism in Korea, see Kim Jae-won, "Riŏllijŭm misul," *Journal of Korean Modern and Contemporary Art History* 22 (December 2011): 71–89.
35 Yi Ch'ŏl-Ho, "Sin Kyŏnghyangp'a pip'yŏng," 235; and Pak Yŏng-hŭi, "Sinhŭng yesul ŭi ironjŏk kŭn'gŏrŭl nonhaya Yŏm Sang-sŏp kun ŭi muji rŭl pakham," *Chosŏn ilbo*, February 3–4, 1926.
36 Yi Ch'ŏl-Ho, "Sin Kyŏnghyangp'a pip'yŏng," 239; and Kim Ki-chin, "Nŏhŭi ŭi yangsim e kobal handa," in *Kim P'al-bong munhak chŏnjip*, ed. Hong Chŏng-sŏn (Seoul: Munhak kwa Chisŏngsa, 1989), 404.

37 Yi Ch'ŏl-Ho, "Sin Kyŏnghyangp'a pip'yŏng," 235–236.
38 Kim Min-chŏng, "Riŏllijŭm ŭi kangbak, chŭngsang ŭrosŏ ŭi riŏllit'i," *Minjok Munhaksa Yŏn'gu* 54 (2014): 315.
39 Yi Ch'ŏl-Ho, "Sin Kyŏnghyangp'a pip'yŏng," 242; Yŏm Sang-sŏp, "Munye wa saenghwal," *Chosŏn mundan*, February 1927; and Han Ki-hyŏng and Yi Hye-ryŏng, eds., *Yŏm Sang-sŏp munjang chŏnjip*, vol. 1 (Seoul: Somyŏng Ch'ulp'an, 2013), 545.
40 Kim Hoe-san, "Misulga ŭi saenghwal ŭisik" (ha), *Chosŏn ilbo*, July 10, 1935. Hoe-san is the pen name of the Western-style painter Kim Chong-t'ae.
41 Kim Min-chŏng, "Riŏllijŭm ŭi kangbak," 316.
42 For details of the Korean political situation after the liberation, see Carter J. Eckert et al., *Korean Old and New: A History* (Seoul: Ilchogak, 1990), 327–346.
43 Im, *Sajin'ga wa ŭi taehwa*, 103.
44 Ibid., 126.
45 Ibid., 103.
46 Koo Wang-sam, "Sajin ŭi riŏllijŭm munje—chakka ŭi iron surip ŭl wihayŏ," *Tonga ilbo*, February 17, 1955.
47 Yi Kyŏng-min, "Charyoro pon Han'guk sajinsa: Haebang konggan ŭi nambukhan sajin'gye ŭi chojik (1)," *Sajin & Munhwa* 1 (February 2010): 5.
48 Yi Kyŏng-min, "Charyoro pon Han'guk sajinsa: Haebang konggan'gwa Han'guk chŏnjaenggi ŭi nambukhan sajin'gye ŭi chojik (2)," *Sajin & Munhwa* 2 (July 2010): 8.
49 Ch'oe Pong-rim, "Che 1 hoe Im Sŏk-che yesul sajin kaeinjŏn gwa kŭ i hu," *Sajin & Munhwa* 9 (July 2015): 2–6.
50 Yi Kyŏng-min, "Charyoro pon Han'guk sajinsa: Haebang gonggan," 9.
51 Yi Kyŏng-min, "Charyoro pon Han'guk sajinsa: Haebang konggan," 9.
52 Yi Kyŏng-min, "Charyoro pon Han'guk sajinsa: Haebang gonggan," 6.
53 Limb Eung-Sik, "Han'guk sajin'gye ŭi hyŏnhwang- chakhwa kyŏngyang ŭl chungsim ŭro," *Sŏul sinmun*, August 1959; and *Im Ŭng-sik hoegorok*, 289.
54 Limb Eung-Sik, "Sajinin ŭi pullyu," *Chŏnmang,* (Ch'anggan ho) (September 1955); and Limb, *Im Ŭng-sik hoegorok*, 291–292.
55 Lim Seok Je (Im Sŏk-che), "Sajin ŭi yesulsŏng e kwanhayŏ," *Tonga ilbo*, October 3, 1954.
56 Koo Wang-sam, "Chuch'e ŭisik hwangnip hara—che7hoe sahyŏpchŏn ŭl pogo," *Tonga ilbo*, November 12, 1955.
57 Limb Eung-Sik, "Kukchŏn e sajinbu rŭl," *Chosŏn Ilbo*, August 30, 1955; and Limb, *Im Ŭng-sik hoegorok*, 247–248.
58 Limb Eung-Sik, "Sajin ŭi yesulsŏng gwa silche," *Democratic Policeman*, March 1956; and Limb, *Im Ŭng-sik hoegorok*, 296.
59 Yi Myŏng-dong, "Sajin ŭi yesulsŏng–Kukchŏn ŭi sajinbu sŏlch'i munje," *Tonga ilbo*, July 27, 1957.

60 Im, *Sajin'ga wa ŭi taehwa*, 98–100.
61 Ibid., 155–158.
62 Ibid., 47–61.
63 Ch'oe In-jin, "Haebang ch'ogi ŭi sajin chapchi yŏn'gu- Kukche Podo wa Sajin Munhwa rŭl chungsim ŭro," *Aura*, no. 10 (2003): 8–17.
64 Ibid., 15.
65 Several photographers including Yi Hyŏng-rok, Chŏng Pŏm-tae, and Han Yŏng-su recollected the critical role of *Sinsŏnhoe* in disseminating Korean realist photography. See the interviews in Im, *Sajin'ga wa ŭi taehwa*.
66 Pak Chu-sŏk, "Haebang hu 1950–1960 nyŏndae han'guk sajin ŭi chŏn'gae wa sŏnggwa," *Han'guk hyŏndae sajin 60 nyŏn, 1948–2008* (Kwach'ŏn: National Museum of Contemporary Art, Korea), 20.
67 Im, *Sajin'ga wa ŭi taehwa*, 77.
68 Ibid., 75.
69 For more on Alfred Stieglitz and the Photo-Secession group, see Ulrich F. Keller, "The Myth of Art Photography: A Sociological Analysis," *History of Photography* 8, no. 4 (October–December 1984): 249–275.
70 Cho Wŏn-myŏng (p'yŏnjippu), "Sinsŏnhoe sajinjŏnp'yŏng—che 1 hoe palp'yojŏn ŭl pogo," *Sajin munhwa*, April 1957, 89.
71 Limb Eung-Sik, "Saenghwal chuŭi ga churyu—1958 nyŏn sadan kyŏlsan," *Segye ilbo*, November 27, 1958; and Limb, *Im Ŭng-sik hoegorok*, 283.
72 Im Yŏng-gyun, *Sajin'ga wa ŭi taehwa*, 78.
73 "Taedam amach'uŏ-ŭi chillo—Yi Hae-sŏn (sayŏn) Im Ŭng-sik (sahyŏp)," *Sajin munhwa*, January 1957, 49–50.
74 "In'gan kajok sajinjŏn," *Sŏul sinmun*, March 28, 1957. The newspaper announcement is as follows: "The photographic exhibition *The Family of Man* that is to be held by USIA was supposed to open from March 25, 1957, but due to the delay of its preparation, it was postponed and will be held for twenty-six days from April 3 to 28, 1957."
75 Limb Eung-Sik, Yi Hyŏng-rok, Ch'oe Min-sik, Yi Kyŏng-mo, and others have mentioned the impact on their photographic practices in the interviews with Im Yŏng-kyun. See Im, *Sajin'ga wa ŭi taehwa*.
76 Museum of Modern Art Archives, New York, Collection: Edward Steichen Archive (ESA), Series folder V. B. ii. 16.
77 Museum of Modern Art Archives, ESA, Series Folder V. B. ii. 16. The number of spectators was based on the statistics from 1955 to 1959. It is known that the exhibition was viewed by more than nine million people in sixty-nine countries.
78 Ibid.
79 Museum of Modern Art Archives, ESA, Series folder V. B. i. 46.

80 Monique Berlier, "The Family of Man: Readings of an Exhibition," in *Picturing the Past: Media, History and Photography*, ed. Bonnie Brennen and Hanno Hardt (Urbana: University of Illinois Press, 1999), 213.

81 Edward Steichen, *The Family of Man* (New York: Museum of Modern Art, 1955), 4–5; and "Photography: Witness and Recorder of Humanity," *Wisconsin Magazine of History* 41, no. 3 (Spring 1958): 159–167.

82 Carl Sandburg, prologue to *The Family of Man* (New York: Museum of Modern Art, 1955), 3.

83 Ibid.

84 For the details and insightful analyses of the exhibition *The Family of Man*, see Bonnie Brennen and Hanno Hardt, ed., *Picturing the Past*; Gerd Hurm, Anke Reitz, and Shamoon Zamir, *The Family of Man Revisited: Photography in a Global Age* (London: I. B. Tauris, 2018); and Eric J. Sandeen, *Picturing an Exhibition: The Family of Man and 1950s America* (Albuquerque: University of New Mexico Press, 1955).

85 Berlier, "Family of Man," 218.

86 Allan Sekula, "Traffic in Photographs," in *Photography Against the Grain: Essays and Photo Works 1973–1983* (Halifax: Press of the Nova Scotia College of Art and Design, 1984), 96.

87 Berlier, "The Family of Man," 223. For details of the USIA, see also Robert Elder, *The Information Machine: The United States Information Agency and American Foreign Policy* (Syracuse, NY: Syracuse University Press, 1968).

88 Walker, "Context as Determinant," 56.

89 "Family of Man ŭl malhanŭn chwadamhoe," *Sajin munhwa,* May 1956, 41–45; P'yŏnjipsil, "In'gan ilga chŏn ŭi sogae," *Sajin munhwa,* May 1956, 45–46; "In'gan kajok chŏnsijŏn," *Sŏul sinmun*, March 28, 1957; and "In'gan kajok sajin chisang chŏnshihoe," *Tonga ilbo*, April 5, 1957.

90 "Family of Man," *Sajin munhwa*, 41. Limb Eung-Sik mentions that Margaret Bourke-White's two photographs of the two Korean weeping women who were holding the coffin and a man bowed to the ground in the sacrificial ritual were excluded in the exhibition held in Korea. A large-format photograph of trees was also excluded. Yi An-sun's photograph of praying aged women in the Buddhist temple shrine was included.

91 Yi Myŏng-dong, "Riŏllijŭm ŭi t'ongil- sajin chakka hyŏphoe kuhoejŏn ŭl pogo," *Tonga ilbo*, November 16, 1957.

92 Kim Yŏng-ju, "Han'guk misul ŭi tangmyŏn kwaje," *Yŏnhap sinmun*, January 24–28, 1957; Kim Yŏng-ju, Hyŏndae misul ŭi panghyang kwa naeyong," *P'yŏnghwa sinmun*, January 19–26, 1957; and Ch'oe Yŏl, *Han'guk hyŏndae*, 160.

93 Ch'oe Yŏl, *Han'guk hyŏndae*, 149, 154.

94 Ibid., 87.

95 Im Yŏng-gyun, *Sajin'ga wa ŭi taehwa*, 128–130.
96 "The Family of Man ŭl pogosŏ—chwadamhoe," *Sajin munhwa,* April 1957, 40–44.
97 Ibid., 43.
98 Im Yŏng-gyun, *Sajin'ga wa ŭi taehwa*, 60.
99 Victor Burgin, ed., *Thinking Photography* (New York: Palgrave Macmillan, 1982), 2.

CHAPTER 9

Long-Distance Recall

Nam June Paik and the Prosthetics of Memory

Steve Choe

Nam June Paik, during his prolific and varied career as an artist, musician, and thinker of electronic media, has striven to realize more capacious, more humane ways of relating to technology. This chapter discusses two installation pieces, *Electronic Superhighway: Continental U.S., Alaska, Hawaii* and *Megatron/Matrix,* both from 1995 and on permanent display in the Smithsonian American Art Museum, to show how the human capacity for retention has been constituted technologically through moving image archiving. *Electronic Superhighway* is made up of fifty-one video feeds, each transmitted to a bank of monitors dedicated to a US state. They feature archival footage drawn from commercial, industrial, and other sources, as well from Paik's previous video work, that showcase both stereotypical and unknown aspects of each state. The 215 monitors that make up *Megatron/ Matrix* display found footage as well, appropriated images that depict hand-drawn cartoon characters, Korean folk dancing, national flags, images from the Seoul Olympics in 1988, and abstract computer graphics, among many other visual representations. I attempt to show how these pieces by Paik seem to model the capacity of modern consciousness to recall the past, to remember histories that have and have not been lived, through communication technologies that function as prostheses to human consciousness.

As scholars have noted, Paik's work has been interpreted to anticipate the incorporation of imaging and archival technologies into the everyday lives of human beings. In this regard, we may consider the ways in which the "electronic superhighway" has already transformed the status and dissemination of knowledge, which in turn has radically altered the criteria for legitimizing knowledge claims themselves. Ready access to the internet on smartphones and mobile devices has trained their human users to put little effort into recalling names, facts, and histories because all knowledge can be quickly retrieved with a simple search. Paik's work provides a critical view on these transformations, showing how the "outsourcing" of memory to electronic images removes ontological differences that have traditionally marked spatial and temporal distinctions between near and far, now and then.

Toward the end of this chapter, I show how the work of this American artist may be read to reflect on the ramifications of this increased complicity between humans and technology in regard to Korean national specificity. After leaving Korea for Japan and West Germany, Paik eventually settled in New York in 1964, becoming a naturalized US citizen in 1976. The artist had lived outside his country of birth for almost four decades by the time *Electronic Superhighway* and *Megatron/Matrix* were produced and during this time observed the turbulent political developments of South Korea, its tense and protracted relationship to the Cold War, and its meteoric economic rise from afar. As we shall see, Paik's later work seems to acknowledge his long absence from the country of his birth while paradoxically recalling Korea as a series of electronic images that have become global, cosmopolitan signifiers.

Paik interest in the relationship between humans and technology is evident early on in his career, in the work he produced soon after arriving in the United States. In collaboration with the engineer Shuya Abe, Paik built an anthropomorphic automaton called *Robot K-456* (1964), named after Mozart's eighteenth piano concerto in B-flat. Constituted of aluminum, wires, old speakers, and hinges barely holding the parts together, the robot looks as if it is about to collapse at any moment. The piece is modeled after a fragile human body rather than the prototypical superbeing of science-fiction whose presence implicitly poses a threat to mortal man. The vulgar gracelessness of *Robot K-456* not only refutes but also cheekily mocks the precious elegance of the Mozart concerto. Yet because of its clumsiness and seeming vulnerability, the robot invites others to mingle and interact with it as it lumbers along city streets. Remotely controlled, it could be commanded to wave, rotate its breasts, bow slightly, tilt its square head, and excrete beans. From various speakers mounted on its body, a recording of JFK's 1961 inauguration is played, mischievously mocking the momentousness of this celebrated president's speech.

In 1982, during a retrospective of Paik's work at the Whitney Museum, *Robot K-456* was rolled down Madison Avenue and, while crossing 75th Street, was struck by a Honda Civic driven by the artist William Anastasi. Recalling a Fluxus Happening, Paik called this staged performance the *First Accident of the Twenty-First Century* (1982) and had footage of it broadcast on a local CBS affiliate. When a reporter asked what it all meant, the artist responded, according to John Hanhardt, "that he was practicing how to cope with the disaster of technology in the twenty-first century. He also noted that the robot was twenty years old and had not had its Bar Mitzvah."[1] In the twentieth century, Paik seemed to anticipate the era of humans encountering technology in the twenty-first, of scooters, food delivery robots, and self-driving cars traversing sidewalks and roads and with it the rising possibility of accidents between organic and inorganic beings. He provokes observers to consider the preconditions for coping with technological disaster and to critically rethink the relationship between intelligent technologies and the human body. Paik's playful *First Catastrophe of the Twenty-First Century* draws attention to the possibility of coexisting with machines, which potentially have the capacity to think, not as ontologically other to the everyday life of humans, but as intentional beings that may somehow be sympathized with. Is it possible to mourn for the fragile, incapacitated robot that has been hit by an automobile, driven by a human who intends to run it down? To what extent are concepts such as guilt, innocence, and debt, concepts that have traditionally revolved around the moral human being, relevant in understanding the ethics of such accidents?

If Paik compels us to think critically about the relationship between humans and technology through the staging of an accidental event, the event of art itself seems to be at issue in his *TV Bra for Living Sculpture* (1969). Constructed from two small video screens, a strip of vinyl, wires, and plexiglass boxes, the piece is a functional brassiere intended for use by his collaborator, the cellist Charlotte Moorman. As Moorman wears the bulky piece, she plays music by Bach and other classical composers on the cello and the two screens covering her breasts show a live television broadcast. The music produced by her cello modulates and disrupts the images shown on the screens. In the performance notes for the *TV Bra,* Paik writes, "The real issue implied in 'Art and Technology' is not to make another scientific toy, but how to *humanize* the technology and the electronic medium, which is progressing rapidly—too rapidly."[2] By bringing new technologies into the tradition of classical music, a tradition that has primarily featured acoustic instruments, Paik seems to be affirming the possibility of thinking video as an expressive medium alongside the cello. Moreover, *TV Bra* allows live broadcast images to be incorporated into the cloistered realm of classical music, music that is often considered nonrepresentational and abstract, thus breaking down boundaries between art and (modern) life. The affirmation of such hybrids and material contaminations seems to be derived from the aesthetic aims that underpin John Cage's *4'33"* (1952), a work that implores listeners to hear noises such as coughing, bodies shuffling in seats, faint vehicle noises outside the concert hall, and other incidental sounds as integral to the performance. In a similar vein, Paik's technological bra, as a kind of found object, is integrated into the cello music that Moorman performs, as the artist writes:

> *TV Brassiere for Living Sculpture* (Charlotte Moorman) is also one sharp example to humanize electronics . . . and technology. By using TV as bra . . . the most intimate belonging of human being, we will demonstrate the human use of technology, and also stimulate viewers NOT for something mean but stimulate their phantasy to look for the new, imaginative and humanistic ways of using our technology.[3]

Both the cello and the TV bra bring about the artistic event and both may be thought of as prosthetic technologies that are incorporated by and extend the capacities of the human body. Commenting on Paik's relationship to Cage, Edith Decker-Phillips writes that though Cage understood pianos, radios, and audio technology as ways of making music, "Paik also saw them as objects with visual qualities."[4] Instead of thinking the classical cello as having a more rarefied status than the small screen that broadcasts local television feeds, both become equally significant in Paik's work for the invocation of the aesthetic experience.

Paik's *Electronic Superhighway* takes the aims of *Robot K-456* and *TV Bra for Living Sculpture* further by collapsing distinctions between high and low art while integrating technology more closely with the potentialities of the body. In doing so, this sculptural video piece shows how our relationship to archiving technologies profoundly informs our everyday knowledge about the United States. On encountering *Electronic Superhighway* on the third floor of the Smithsonian American Art Museum, one is immediately struck by the massive scale of the installation that affects the viewer regardless of whether the piece is seen up close or from afar.

Fifteen feet high, it consists of separate video channels, one for each state plus an additional channel for the District of Columbia, transmitted to 336 television screens mounted on a large scaffold. In front of the bank of screens are a series of neon lights that function as an overlay, delineating the borders of the states and allowing the totality of the map to appear. Indeed, Paik's piece raises a number of ontological questions already introduced by Jasper Johns's painting *Map* (1962)—is it a found object, symbol, simply a sign, or perhaps all three at once?—but heightens the tenuous ontological status of art with history with the introduction of mesmerizingly distractive video images. The content of each video feed is a series of banal images as they would be known nationally or even globally: the Iowa caucuses, New York's Empire State Building, images of potatoes for Idaho. For three states, Missouri, Kansas, and Mississippi, clips from three well-known films are featured: *Meet Me in St. Louis* (1944), *The Wizard of Oz* (1939), and *Show Boat* (1951). Their titles are explicitly labeled with a wooden sign.

As a person approaches *Electronic Superhighway,* they may attempt to overcome the experience of immersion induced by its considerable size and focus on a specific state to see what images Paik has chosen to feature. The state of California, for example, features a quickly changing montage of

Fig. 8 The archive of American memory. Photo by the author.

images of the Golden Gate Bridge and O. J. Simpson leading a fitness aerobics class, among others. The viewer may shift their focus toward a state where they have lived or visited, such that the images, say, of the Kentucky Derby remind the spectator of a past moment when they were physically there. Or they may focus on states where they have never been, such as Oregon, whose screens feature images of mountains, outdoor sports, and the phone number of the state's tourism commission. Audio speakers, embedded between the monitors, carry the sounds derived from a few of the image feeds including Martin Luther King Jr.'s speech in Selma from 1965 and the soundtracks to the Hollywood films just mentioned.

Fig. 9 Seeing oneself seeing. Photo by the author.

The visitor to the exhibit may then turn toward the small screens that make up the feed from Washington, DC, which features a closed-circuit transmission from a camera pointed at the viewer. Standing in the Smithsonian museum in front of Paik's work, in a gesture toward intimate proximity and self-reflexivity, thus recalling Vito Acconci's video work, viewers see themselves seeing as they approach these monitors. For all the other states physically distant from the viewer, he or she may come to realize that the contents of all fifty feeds have been derived largely from the archives of film and television history, the collective memory that constitutes the imagined community of US national identity.

A few states feature footage of Paik's friends and collaborators, such as John Cage in Massachusetts, Charlotte Moorman in Arkansas (she was born in Little Rock), and Merce Cunningham in Washington State. They reflect, perhaps, not so much the memories and associations of the US populace, but those of Paik himself, appropriating extended clips from his previous work. The images for Washington State are derived from Paik's 1978 single-channel video piece, *Merce by Merce by Paik.* In it, Cunningham's image is doubled and tripled and his body is composited on artificial and natural backdrops. Nevertheless, given the copious number of monitors and Paik's penchant to combine high art and popular culture in bewildering montages, it is difficult at first to ascertain how to approach *Electronic Superhighway.* Standing back from it, the effect of the map's entirety is overwhelming, even sublime, but when one steps forward to examine the details of a particular channel feed, the effect is again engulfing because most images appear so quickly, shift perspectives so rapidly, that the viewer is unable to fully take them in. In each, Paik showcases a frenetic mélange of images that combine animation with live action, documentary footage with fictional elements, avant-garde and commercial source materials through a series of colorization and compositing techniques.

Just as the content of the feeds are derived from the archive of popular and Paik's individual memory, so does their disorienting form have a precedence in the artist's oeuvre. We might draw our attention in this regard to another work by Paik called *Global Groove,* a thirty-minute, single-channel video piece from 1973. It begins with a solemn prediction about the future of video and broadcast television that sets the tone for what will follow: "This is a glimpse of a video landscape of tomorrow when you will be able to switch on any TV station on the earth and TV guides will be as fat as the Manhattan telephone book." It then continues with a dizzying montage of archival footage from high and popular art sources, similar in this regard to *Electronic Superhighway,* of performances by Cage and Allen Ginsburg and the Living Theater, commercials from Japanese television, and heavily manipulated footage of a dancing couple. An image featuring the head of Richard Nixon is psychedelically manipulated with a large magnet, recalling Paik's "prepared televisions," the first of which was conceived in 1965 with his found sculpture, *Magnet TV.* Like *Robot K-456, Global Groove* seems to be put together almost haphazardly with discarded materials and awkward combinations of sounds and images. The video's nondiegetic soundtrack blasts Mitch Ryder's "Devil with a Blue Dress On," which is juxtaposed with spoken word texts by Cage, Moorman, and Paik. *Global Groove* celebrates a brave new world in which television viewers are bombarded with a nonlinear series of images and sounds that appear to the viewer as if he or she were switching quickly between countless channels. Alternatively, it is as if Paik's video itself were a manifestation of a kind of media consciousness that brings together disparate cultures and communities through their visual synchronization. In a 1974 review that appeared in the New York Times, Hilton Kramer writes of the rapid rhythms that *Global Groove* foregrounds in regard to the video medium:

> What is important, above all, in this medium is the pace at which the screen projects and devours its images. It is a pace that deliberately subverts any empathetic response we may bring to a specific image, for no matter how compelling—or boring—a particular moment may be, it is "cut" to a rhythm that negates our interest in it in order to fasten all attention on the rhythm itself. This is, in other words, a medium in which representational images are used for the purposes of kinetic abstraction.[5]

The wild graphic manipulations, quick cuts, and jarring juxtapositions featured in *Global Groove* flash by quickly calling attention not to individual signs and symbols but to their torrential onrush through the experience of mesmerizing commotion.

Electronic Superhighway clearly draws from the dizzying form of *Global Groove,* multiplying its single channel by fifty across a map of the continental United States. Through this array of screens, one's attention on any one area or moment of the piece constantly shifts because the montage of images that make up a single state compels not attentive contemplation, but instead a highly distracted form of spectatorship. Walter Benjamin famously wrote of concentration and distraction to describe the consumption of art in modernity and noted that they "form polar opposites which may be stated as follows: A man who concentrates before a work of art is absorbed by it. . . . In contrast, the distracted mass absorbs the work of art."[6] Indeed, Paik's *Electronic Superhighway* appeals, particularly in the almost instinctual recognizability of its imagery, to the preoccupied museum visitor who nevertheless seeks both engagement and immediacy. Experience of the installation may be overwhelming, yet the piece does not compel the viewer to adapt to its aura of aesthetic experience. It addresses attendees through its haphazard form, which has nevertheless become increasingly familiar in a media environment offering countless channels, each of which demand individual attention. Whether *Electronic Superhighway* is seen from afar, where a viewer can experience the ceaseless chaos of its fifty-one video feeds, or from up close, where a person's attention on a single feed constantly shifts in accordance with its rapid rhythm and kinetic abstraction, Paik's piece appeals to a collective archive of images, loose associations, and radically disjointed memories that constitute the idea of America.

In its appropriation of archival sounds and images, *Electronic Superhighway* may remind us of what philosopher Bernard Stiegler calls "tertiary memory." Stiegler's account of tertiary retention is perhaps his most important contribution to the phenomenology of time and, ultimately, to what he calls the "invention" of the human. It is a form of recall that has become increasingly crucial in our age of cinema and the internet and describes how we can recall events that have not been experienced in real life. These technologies of memory and archiving have nevertheless come to function unproblematically as artificial extensions of the capacity of the human being to remember the experiences of others. In volume three of his sprawling work *Technics and Time,* he writes that "tertiary retention is in the most general sense the prosthesis of consciousness without which there could be no mind, no recall, no memory of a past that one has nor personally lived, no culture."[7] A viewer may have never attended the Kentucky Derby, seen O. J. Simpson's aerobics class, or been in Selma to hear Martin Luther King Jr. speak in 1965, but the images of these events recorded onto film and video allow a person to see them take place in the present moment, yet from a temporal and spatial distance.

This capacity to bring closer things that are far from the viewer is perhaps obvious but Stiegler makes the observation more meaningful by extending it to other forms of "technological media," including material artifacts such as clay pots, arrowheads, and paintings, as well as clothing, scientific instruments, and old televisions. Appropriating insights offered by Gilbert Simondon, Stiegler notes that each of these objects implies historical worlds and discursive contexts in which these objects were put to use by past human consciousnesses and bodies. Extending this phenomenology further, the notion of writing itself, and the entire scientific-conceptual edifice depends on it, may be thought of as a type of media as well. If technologies such as the bicycle and the automobile extend the capacity of the human body to conquer space, recording technologies such as film and video extend the capacity of human consciousness to remember.

Yet by becoming habituated to the use of such technologies, the activities and gestures of the body change throughout human history as consciousness adapts to accommodate the capacities opened up by prosthetic memory. These transformations in turn are coopted by the very notion of the human, a notion that is constantly undergoing revision yet in a manner that maintains its distinction from nonliving machines. "The evolution of the 'prosthesis,'" Stiegler writes in reference to this distinction, "not itself living, by which the human is nonetheless defined as a living being, constitutes the reality of the human's evolution, as if, with it, the history of a living being characterized in its forms of life by the nonliving—or by the traces that its life leaves in the nonliving."[8] The human does not simply invent the tool, but the tool also invents the human. Each of these prostheses enable human life to perform specific actions, extending their capacity to colonize and control the world that surrounds them while reiterating and consolidating once again the very idea, indeed the very technics, of the human. Stiegler traces the genealogy of this consolidation back to the myth of Prometheus and Epimetheus in volume one of *Technics and Time,* to the "fault" of Epimetheus and the forgetting of the question of Being (as formulated by Martin Heidegger in his critique of metaphysics).

As a technics manifest in the merging of technological and human capacities, tertiary retention exists as a supplement to primary and secondary retention. Stiegler draws on Edmund Husserl's phenomenological analysis of a melody unfolding in time and these two forms of retention it compels. As a temporal object, a melody is perceived in the immediate moment following the act of hearing, when hearing emerges into aesthetic experience, or when immediate perception develops into past reflection.[9] Primary retention takes place at precisely the moment that the now passes into the just-now, eventually becoming a then. It is constituted as a perceptual unity whose leading edge is embodied in the current note perceived by consciousness and which carves out and shapes time from a future that is as yet undifferentiated. In addition, secondary retention names the capacity to recall an event at a much later time, the act of recalling a moment that already took place but is no longer present to perception. When a person sings a melody heard the day before, the act calls on the individual to reconstruct the temporal object seemingly out of thin air, to experience it again through its reconstruction. The singing of a familiar melody is an act of remembering and a re-embodiment of the past that also takes place in the here and now.

Because the phenomenological method disregards all phenomena that cannot be directly experienced by consciousness, Husserl is unable to describe the experience of tertiary retention beyond its primary and secondary forms. Tertiary retention remains unique in that its existence depends on recording technologies such as writing, film, and video. By representing past moments that have not been lived personally, these technologies give credence to the existence of histories, of cultures, and of times and places that have not been lived. Stiegler's most interesting insight into the phenomenon of memory is to show that, in the age of mechanical reproducibility, the strict division between primary and secondary retention delineated by Husserl dissolves and is shown to be subsumed in tertiary memory. As a technics of historicity, or as a form archi-writing that constitutes the very notion of pastness, the act of recall is deeply informed by mechanical reproduction such that the meaning of pastness and of recall themselves are always already technological. Consciousness, Stiegler writes, "is always in some fashion a montage of overlapping primary, secondary, and tertiary memories."[10] And memory, he goes on to note,

> in all its forms would then always be a sort of rushing montage of frozen images, from the simplest juxtaposition to the greatest art of the scenarist, according to the quality of the consciousness and the nature of the object presented to it, and according to the criteria—the secondary memories, i.e., the experiences—it evokes from the object.[11]

The act of recall in the age of technological media, in other words, shuttles between primary and secondary retention. Yet the sense that both exist, in their originary form, is only possible because of the fundamental structure constituted by tertiary retention.

To demonstrate how this is the case, Stiegler critically develops Husserl's phenomenology of a melody unfolding in time to include the consideration of technological media. The initial hearing of a melody seems to compel a kind of description proposed by the German philosopher, but the repeated hearing of the same melody recorded on technological media, on what he calls a "phonographic support mechanism," introduces new phenomenological variables in regard to the experience of time and temporality. After the second hearing of the melody, the listening consciousness has already and irreversibly changed. One knows how the melody will be phrased, when it will culminate, and how it will resolve at its conclusion. Explaining this, Stiegler writes,

> From one hearing to another it is a matter of different ears, precisely because the ear involved in the second hearing has been affected by the first. The same melody, but not the same ears nor, thus, the same consciousness: consciousness has changed ears, having experienced the *event* of the melody's first hearing.[12]

If a person claims to have experienced the melody differently on the second hearing, the difference is accounted for by the fact that perception, as Husserl points out, is always and necessarily selective. Yet as the melody is heard again, new details may be discovered and through this it becomes more deeply inscribed in memory, its trajectory from beginning to end becomes more deeply etched into consciousness and the gestures of the body. Secondary retentions inhabit the primary experience on the second hearing and the experience of listening becomes less the discovery of a melody never heard and increasingly an expectation of its particular phrasing, culmination, and resolution. The individual knows in advance that it is being heard again, that the listening experience to come will be inspired by its initial hearing. On the other hand, Stiegler states that, in fact, this way in which secondary retention always already inhabits primary retention "is also the case when I have never heard it, since then I hear from the position of an expectation formed from everything that has already musically happened to me—I am responding to the Muses guarding the default-of-origin of my desire, within me."[13] The experience of a recorded melody, even on its first hearing, is predetermined by the system of harmony, rhythm, the idea of a melody, and the act of listening itself—themselves all forms of tertiary retention. Consciousness, which already knows what a melody is, itself undergoes critique through this proactive consciousness. As recording technologies increasingly determine the experience of the temporal object, so the concept of consciousness, itself a temporal object, becomes increasingly determined by the technics of recording. Human consciousness itself becomes not a living entity separate from the presumptively nonliving melody but instead one with it, synchronized with its flow. It becomes unclear where the organic human ends and an inorganic technics begins. Yet, once again, so that consciousness may identify itself as a specifically human consciousness that hears and aesthetically appreciates the melody, it must disavow the fact that the very concept of the human is already technical, constituted through reiteration, and thus must forget that the possibilities of experience for primary and secondary retention are already severely delimited.

Stiegler's insights into tertiary retention help explain the overwhelming, yet seemingly mundane, experience of *Electronic Superhighway.* If this experience may be characterized as radical distraction, it is because the rapid montage of images appeals, not in a manner that invites contemplative appreciation, but that allows itself to be unproblematically "absorbed" by the observer. Familiar images flash before the person, but their brief duration forecloses thoughtful consideration and careful critique. The capacity to recognize these images, such as of Florida orange groves or Iowa cornfields, is directly attributable to the archive of tertiary retention. This prosthetic archive consists of images and sounds that make up the universally understood image of Florida or Iowa, the mundane images that inform what may be known about each US state. That a person can recognize this way, regardless of whether they have experienced the orange groves or the cornfields in these locations, attests to the power and discursive reach of tertiary retention. In this, the capacity to quickly recognize the images presented in Paik's work, which is already enough to understand and appreciate *Electronic Superhighway,* is concomitant with the extent to which a person has absorbed the films, television, writings, and other forms of tertiary memory that reiterate images of (in these examples) Florida and Iowa. That these feeds are looped and repeat endlessly seems to underscore the importance of this form of technological retention.

Moreover, *Electronic Superhighway* seems to confirm the extent to which thinking the totality of the United States of America is made possible only by a consciousness that is always already technical. For beyond the question of whether presenting a large map of the United States as art is a sign, symbol, representation, or simply a found object, Paik extends the scope of this question initially raised by Johns to illustrate how national consciousness is itself formed through tertiary retention. Even though one may have never been to North Dakota or Alabama, the act of recognizing the montages Paik presents, standing in for those states, indicates that one somehow knows these unknown places and thus draws from a base of knowledge that is perceived to originate from nowhere. Indeed, this unlocatable and forgotten archive embodies the myth of America and constitutes a legacy that is bestowed onto us as seemingly unmediated. A form of knowledge that is underpinned by tertiary memory through and through, yet one that is creative and full of possibility, was perhaps envisaged by Paik when he first introduced the term "electronic superhighway" in 1974 in a proposal he wrote for the Rockefeller Foundation. Paik writes that communication networks spanning the United States "will cease to be an Ersatz (substitute) or lubricant but will become the springboard of unexpected new human activities."[14] In the exhibition catalog for his show at the 1993 Venice Biennale, Paik claimed that "Bill Clinton stole my idea," specifically regarding the possibility of connecting "New York and Los Angeles with multi-layer of broadband communications networks," thus foreseeing the emergence of the internet.[15] Today, a person might reflect on the extent to which tertiary retentional technologies such as YouTube, Spotify, and Netflix have quickly come to dominate how we experience temporal objects, melodies, and moving images. *Electronic Superhighway* seems to explore how the logic of the network has given rise to new human activities and possibilities, specifically the ways in which human memory has been expanded through the internet as the latest, and now most important, prosthetic of consciousness.

If *Electronic Superhighway* concerns the network of moving images and sounds that constitute the totality of American identity, *Megatron/Matrix* attempts to imagine a network that contains particular, local images (evoking the particularity of South Korea) yet also belong to an archive of tertiary memory that is cosmopolitan and global. Its scope is nearly as monumental as *Electronic Superhighway,* standing eleven feet high and thirty-three feet wide, featuring two large blocks of screens juxtaposed to each other: one rectangular group of fifteen by ten monitors on the left ("Megatron") and another square of eight by eight monitors on the right ("Matrix"). Like the kinetic montages featured in *Global Groove,* Megatron is made up of a series of images shown in a loop of about twenty minutes in duration. They include hand-drawn animals and characters appropriated from Korean animation, footage from *Merce by Merce by Paik,* live action video composited from layers of computer graphics, snippets of athletes playing various sports from the 1988 Seoul Olympics, Charlotte Moorman playing the cello, and animated clips of traditional Korean folk dancing. Although each screen may feature a single video channel, most of these images are constituted and move across the entire bank of monitors, creating a spectacular effect whereby each screen, each like a pixelated element, contributes to generating images that are larger than life. As is true of Paik's other video montages, fast cutting, kineticism, and the gestural body are underscored, producing a mesmerizing effect on the viewer. The frenetic music that accompanies the images is partially adopted from the soundtrack to *Global Groove* but adds references to traditional Korean drumming and modern T'ŭrot'ŭ pop music.

Most striking are the series of national flags that periodically materialize and appear on the Megatron screens. Countries such as Canada, Germany, Italy, Sweden, and Korea are represented, recalling the athletic competition between nations that takes place during the Olympic Games. Their appearance seems to be expressing the hope that Korea could join other cosmopolitan industrialized nations and be included in the repertoire of world cultures. At the time *Megatron/Matrix* was produced, South Korea was struggling toward modernity and democratization yet was beginning to emerge as a potential economic player on the global market. Paik was certainly aware of how the rise of Korean *chaebŏl* corporatism and events such

Fig. 10 Canada and Merce. Photo by the author.

as the Seoul Summer Olympics would raise the country's profile in the world. In 1988, the year the Olympics took place as a global media event, he created *Dada-ikseon (The More, the Better)*, a towering, sixty-foot tall installation piece consisting of 1,003 televisions that is now featured at the National Museum of Contemporary Art in Seoul. Further, he organized a global broadcast of a forty-seven-minute video piece called *Wrap Around the World* (1988), which was estimated to have been seen by fifty million viewers in more than ten countries. Taking the aesthetics and scope of *Global Groove* even further, the video features Brazilian dancing, Kung Fu demonstrations, and an orchestral performance of music by Brahms, in addition to performances by David Bowie, Ryuichi Sakamoto, and Merce Cunningham.

The imagery featured in the smaller Matrix section of his 1995 piece is loosely related to that in Megatron, but seems to pursue other, perhaps more abstract, aims. The TV monitors are positioned so that they form a swirl around the center, and the images that appear on them emphasize patterns of geometric movement rather than presenting recognizable representations. The screens are coordinated so that they change in kaleidoscopic arrangements, recalling what Siegfried Kracauer calls the "mass ornament." Like the synchronized legs of the dancing Tiller Girls referenced throughout his 1927 essay, the patterns produced by the synchronized screens of Matrix "consist of lines and circles like those found in textbooks on Euclidean geometry, and also incorporate the elementary components of physics, such as waves and spirals."[16] For Kracauer, the mass ornament is an end in itself and represents a limit of capitalist rationality and aesthetics, while carrying the potential to betray the violence of its totalizing logic. The shapes and vague human figures represented in Matrix culminate, not in a meaning that must be uncovered through sustained interpretation, but in a critique of late capitalism that is made discernible to the attuned viewer.

Kracauer writes that the patterns delineated in the mass ornament are devoid of sexuality—"they are a linear system that no longer has any erotic meaning but at best points to the locus of the erotic."[17] At the center of the vortex of the Matrix section is a single screen that shows a looped clip of two partially nude women interacting with each other. The hermeneutically empty patterns delineated by the surrounding screens seem to rotate around this "locus of the erotic" depicting soft-core porn. With its representation of sexuality at the center, one might read Matrix as representing the unconscious to the conscious psyche depicted in the Megatron section. One might thus interpret the wild mélange of signs presented in Megatron as expressions of repressed, illicit desires. Although such readings reinforce the notion that all the signifiers of consciousness point to the truth of sexuality, this psychoanalytic insight gives way only to another signifier in *Matrix/Megatron,* another video image.

For our purposes, the questions of nationhood and belonging in the age of tertiary memory persist and reiterate themselves as in the looped video of this video piece. What does it mean for Nam June Paik to feature imagery and the flag of his birth nation in *Megatron/Matrix* after four decades of not living there, after observing its political turmoil and economic rise from afar through technological media? How does the national particularity and ethnic specificity of Korea become part of universal humanity in the age of tertiary memory? Indeed, is this aim of becoming universal one that should be desired at all?

To address these questions, I return briefly to Stiegler, who interrogates the legitimacy of saying I and We against the backdrop of an increasingly hegemonic and thus universalizing retentional structure determined by technological media. He begins with the premise that "An *I* claiming to make rational and universalizable statements would always be able to say 'we,' and this is precisely what *We* do in this context, asking 'us' whom 'we' are speaking of and in the name of *what* or *whom* 'we' allow ourselves to speak in 'its/her/his' name."[18] To say I means, as Stiegler reminds us, to implicitly adopt a metaphysics of universal humanity by speaking "rationally." In turn, this metaphysics legitimates the capacity for one who says I to also claim the We and to speak on behalf of other speaking humans in a universal manner. Yet a critical question arises from these Kantian premises: what are the epistemological conditions under which this ethical claim toward universality may be legitimized? Who is allowed to be rational and speak on behalf of the We? For Stiegler, language is always already constituted through an archive of memories that were not in fact experienced by individual consciousness such that we are born into a world of already existing languages, cultures, and civilizational schemas. This technics of worlding implicates consciousness in a cycle of repetition and expectation, of pastness and futurity, the one determining the other throughout the lifespan of the finite human. Such a technics of the self can be extended, moreover, to describe the technics of the community more broadly. A community of human beings, habituated to technical consciousness, may be defined in the way it attunes itself toward a shared communal duty, embodied in the concept of nationhood or a civilizational form. To speak on behalf of such a community, one that can say We, means then to identify and take responsibility for its past and for its configurations yet to come. "The unification process of a *We,*" Stiegler writes, "is an identification, an organization, and a unification of diverse elements of the community's past as they project its future."[19] A person belongs to a community by identifying with its technics, embodying the temporality it inaugurates on behalf of its people, and thus taking responsibility for its destiny.

As we have seen, this retentional structure is always already one with the technics of film and video, both mediums of time that unfold and are experienced as temporal phenomena. Temporal objects organize the experience of time for the viewer, the one who says I, whose time-consciousness is synchronized with a global We through the experience of technological media. It shows us how to remember through repetition, reminding us of the human that we ostensibly are, and teaches us how to portend and confront the future. In no other time in human history has the presence of the temporal object, wirelessly streamed to smartphones, tablets, and computers, been more ubiquitous than today. Yet because of this condition, in an age dominated by the electronic highway of the internet, the critique of technics has become increasingly urgent, a critique that Paik's work makes visible to us.

Both of Paik's video pieces discussed here point to the importance of tertiary retention in thinking through the relationship between humans and technology. Yet *Megatron/Matrix* should be understood as his attempt to speak on behalf of the universal, beyond his national-ethnic specificity as an Asian, Korean, American, or even Korean American artist. It thus addresses an epistemological problem posed by Naoki Sakai in a recent essay on the critique of Asian-ness within our current world order: "The Asians may well produce certain wisdom, but their wisdom could never transcend their ethnic particularity and thus reach the domain of theoretical universality."[20] *Megatron/Matrix* attempts to overcome this discursive impossibility by exploring the possibilities of the video medium and the ways it has profoundly altered the relationship of humans to technology. It provokes the thought of whether archival images of Korea and Korean-ness can become part of the repertoire of the tertiary memory of universal humanity, whether these images can become as familiar as Florida orange groves or Iowa cornfields for a global audience. In doing so, Paik's work enables us to consider how the moving image remains fundamental to thinking national particularity as simultaneously a claim toward universality.

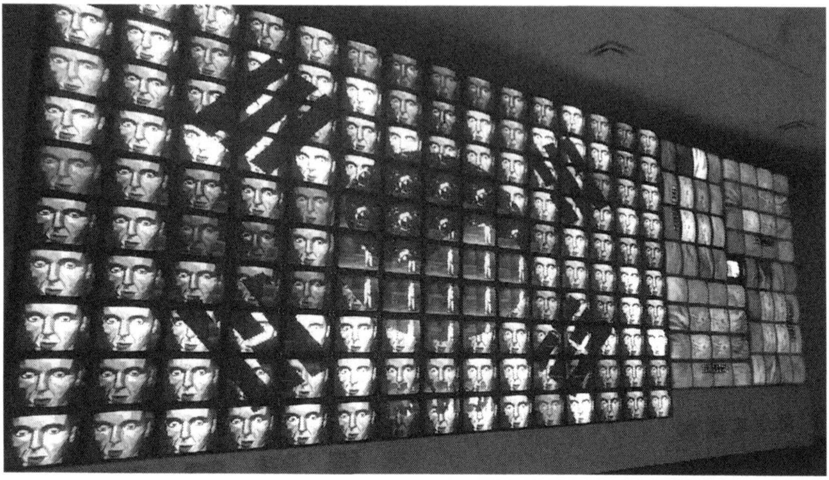

Fig. 11 Korea as technological memory. Photo by the author.

As a kind of consciousness constituted through video images, *Megatron/Matrix* calls attention to how nationality itself exists merely as a signifier within a global archive of national identities, an archive that is grounded in the technics of film and video. What counts as rational and universalizable today must adhere to this logic, one that diverges in many ways from the aims and scope of Enlightenment reason but is constituted in the reproducibility, exchangeability, and manipulability of electronic moving images. This logic is essentially quantifying and objectifying, concerned less with the uniqueness and historicity of the work of art and more with its capacity to circulate and be recognized by audiences who live in South Korea as well as by those who have never visited the country. *Megatron/Matrix* seems to suggest that what counts as national specificity within our technical horizon, and in the name of universal humanity, conforms to the discursive patterns set out by the logic of tertiary retention. The only way in which Korean specificity may thus be universalizable today is in the form of moving image technology and the habits of tertiary memory it inaugurates. Indeed, we recognize this form whenever we speak of the soft power that K-Pop, K-Dramas, and other popular images and sounds of Korea represent to the world. That we can recognize them as such indicates the extent to which these images, as temporal objects, have been accommodated to the habits and patterns of global tertiary memory. As I have tried to show through Stiegler's comments, this form has always already constituted the metaphysics of the community and of the We that names it.

By creating *Megatron/Matrix,* Paik reminds himself what it means to be Korean from afar. It gestures toward an identity that is always already implicated within tertiary retention, an identity that is both particular and yet speaks on behalf of universal humanity through the medium of video. Although he had not lived in Korea for forty years and had been an American citizen for twenty, *Megatron/Matrix* expresses the experience of memory of being Korean and identifying with its temporality, as an act of long-distance recall.

Notes

1 John G. Hanhardt, *Nam June Paik: Global Visionary* (London: Giles, 2012), 36.

2 Nam June Paik, *Videa 'n' Videology 1959–1973,* ed. Judson Rosebush (Syracuse, NY: Everson Museum of Art, 1974), n.p.

3 Ibid.

4 Edith Decker-Phillips, *Paik Video,* trans. Marie-Genviève Iselin, Karin Koppensteiner, and George Quasha (New York: Barrytown, 1998), 31.

5 Hilton Kramer, "A Fast Sequence of Forms Changing Color and Shape," *New York Times,* February 3, 1974, reprinted in *eine DATA base,* ed. Klaus Bussmann and Florian Matzner (Berlin: Edition Cantz, 1993), 225.

6 Walter Benjamin, "The Work of Art in the Age of Mechanical Reproduction," in *Illuminations: Essays and Reflections,* ed. Hannah Arendt, trans. Harry Zohn (New York: Schocken Books, 1988), 239.

7 Bernard Stiegler, *Technics and Time, 3: Cinematic Time and the Question of Malaise,* trans. Stephen Barker (Stanford, CA: Stanford University Press, 2011), 39.

8 Bernard Stiegler, *Technics and Time, 1: The Fault of Epimetheus,* trans. Richard Beardsworth and George Collins (Stanford, CA: Stanford University Press, 1998), 50.

9 Edmund Husserl, *On the Phenomenology of the Consciousness of Internal Time (1893–1917),* trans. John Barnett Brough (Dordrecht: Kluwer Academic Publishers, 1991).

10 Stiegler, *Technics and Time,* 3, 28.

11 Ibid., 28.

12 Ibid., 17.

13 Ibid., 19.

14 Nam June Paik, "Venice IV—1993, Bill Clinton Stole My Idea," in Bussman and Matzner, *eine DATA base,* 110.

15 Ibid.

16 Siegfried Kracauer, "The Mass Ornamant," in *The Mass Ornament: Weimar Essays,* ed. Thomas Y. Levin (Cambridge, MA: Harvard University Press, 1995), 77–78.
17 Ibid., 77.
18 Stiegler, *Technics and Time,* 3, 88.
19 Ibid., 93.
20 Naoki Sakai, "Theory and Asian Humanity: On the Question of *Humanitas* and *Anthropos,*" *Postcolonial Studies* 13, no. 4 (2010): 452.

CHAPTER 10

Affect in the End of Days

South Korean Science Fiction Cinema, *Doomsday Book*, and Affective Estrangement

Haerin Shin

Doomsday Book, a collection of three short films released in 2012, occupies a unique position within the recently emergent cache of science fiction (SF) narratives in South Korean cinema. Instead of keeping the rhetoric of the unreal in check by framing it in the realm of the symbolic or leaning toward social allegory, as other works of its kind in South Korea often do, *Doomsday Book* takes a head-on approach to the genre by foregrounding technology itself as a key player in offering a colorful spectrum of themes that populate the apocalyptic imaginary. SF has remained in the peripheries of Korea's cultural production since its modern turn in the late 1800s as urgent demands to address the incessant upheavals in the sociopolitical sphere continued to preoccupy the literary conscious.[1] Although the privileged position of critical realism continues to assert its dominance, South Korea has grown into a leading figure in the global telepresence technology industry, cutting-edge gadgets and advanced networked infrastructure woven into the minutest aspects of the lifestyle domain, resulting in a curious lag between the instantiation and representation of technology.[2] *Doomsday Book* arrives at a timely juncture to address this gap, situating South Korea's relatively insulated SF scene in conversation with the global trend by addressing a topic that has returned, after a period of latency since Hiroshima, to captivate the world—the apocalypse.

Doomsday Book's end-of-day scenarios address the contemporaneity of the apocalypse narrative with its three-pronged approach. Twisting some of the most popular motifs in apocalyptic and post-apocalyptic fiction into its Mobius strip–like collation of seemingly discrete yet closely interlinked worlds of impending doom, the film situates affective reconfigurations of anthropocentric ontology at the seismic origin of a pollution-driven zombie epidemic, technological singularity, and cosmic disaster. Technologically induced, apocalypse or its prospect serves more as a reconfiguration than a termination of the existent system. For instance, the body's physical capacity to affect mutual attraction becomes the sole proof of life and personhood on the zombie-infested streets of "A Brave New World," where neither a sentient I nor a cognizant Other remains to corroborate what it means to be (consciously alive, in this case). Digressing from the discursive framework of progress and knowledge that sustain the world-building mechanism of modern SF, whether it be from a critical or more celebratory stance, the stories in *Doomsday Book* neither affirm nor vilify the forces that radically reconfigure the human condition but instead acknowledge them as capacities to be lived out without imposing value-weighted hierarchies upon their instantiations. In this light, *Doomsday Book* aptly captures the postmillennial zeitgeist of South Korea's cultural imaginary in the wake of fallen grand narratives such as ideological conflicts between the North and South or the urgency of economic progress that drove South Korea's modernization.[3]

By placing emphasis on affirming vitality regardless of its expression (desire, physicality at large, or the pure drive to survive), *Doomsday Book* reclaims new ground in South Korean cinema. Inviting a novel way of understanding what it means to live and survive in an age when incomprehensive, uncontrollable, and unintended technological developments abound, the film signals what Patricia Clough calls "the affective turn." Affect, according to Clough, means the "bodily capacities to affect and be affected or the augmentation or diminution of a body's capacity to act, to engage, and to connect, such that autoaffection is linked to the self-feeling of being alive—that is, aliveness or vitality."[4] More a force of movement than conscious acts of cognition or responsive perception, physical yet also immaterial in their resistance to formality, affective dynamics destabilize rigid dichotomies that have long sustained the agenda of the modern around the globe under the catchphrase of progress, and acknowledge the present without fettering it to a causal chain of hierarchized temporality.

Beginning with a brief synopsis and reflections on the film's critical and commercial reception, I offer a theoretical framework for interrogating how *Doomsday Book* demarcates a refreshing, affective turn in South Korean SF cinema by portraying how nonhuman agency signals the affective emergence of alternative modes of being, either in the throes of apocalypse or through attempts to process its aftermath. *Doomsday Book*'s approach to representation suggests a way out of the constraints of critical realism as the dominant tradition in modern Korean culture through the narrative schemes of SF while amplifying the critical capacity of the genre to estrange, the object of which is not only reality and modes of its cognition but also the very tenet of reality and its problems, which in South Korea's case are closely tied to modernization and its challenges. The close reading section that follows focuses exclusively on "A Brave New World" instead of addressing all three episodes, partly given the limited space but also because the biomedical trope of the zombie perfectly aligns with the broad theme of the volume.[5] Clough's assertion that "technoscientific experimentation with affect not only traverses the opposition of the organic and the nonorganic; it also inserts the technical into felt vitality, the felt aliveness given in the preindividual bodily capacities to act, engage and connect" is instructive in seeing how *Doomsday Book* inspires alternative readings of what it means to be alive within and beyond the human substrate, and what is more, *how* to be alive, in the face of its imminent truncation.[6]

From Cognitive to Affective Estrangement: Beyond Modern SF

Doomsday Book opens up with "A Brave New World" ("Mŏtchin sin segye"), wherein a mysterious contagion plunges people into a comatose state of zombie-like rave of violence. The cycle sparks into action with research scientist Yun Sŏg-u, who finds a rotten apple in the fridge and tosses it into the food trash bin while carrying out the unsavory task of disposing of household waste. The molding organism proceeds to travel through the recycling system and ends up in mixed livestock feed along with the rest of the sludge, setting off a citywide pandemic. The second episode, titled "The Heavenly Creature" ("Ch'ŏnsang ŭi p'ijomul"), takes place in a near future run on the labor of artificially intelligent robots. When a service robot called Inmyeong begins to exhibit signs of being enlightened at a Buddhist temple, one of the monks summons a technician from the manufacturer to determine the exact nature of this unforeseen development, triggering a fearsome debate between the robot's advocates, the manufacturer, and the master priest of the temple over the integrity of Inmyeong's claim to spiritual transcendence. "Happy Birthday" ("Haep'i pŏsŭde'i") revolves around a little girl who orders an eight ball from an obscure online commerce site to replace the one she lost from her father's billiard set. Instead of the expected parcel, she receives her order in the form of a giant eight-ball-shaped meteor, which crashes into Earth and obliterates civilization. Surviving the impact by hiding out in a bomb shelter, the family ventures out a decade later to encounter the mysterious seller who descends on the ruins from an extraterrestrial world to confirm delivery.

Doomsday Book, then, is a three-part collection comprising loosely tied stories that converge on the theme of apocalypse. Production began in 2006, only to encounter a prolonged hiatus due to budgetary issues. The finished product was finally released in 2012 to the heightened expectations of South Korea's SF aficionados. The general public also appears to have been drawn to this rare breed of Korean cinema boasting a cast of well-respected actors such as Yu Sŭng-bŏm, Pak Hae-il, Pae Tu-na, and even director Bong Joon-ho (who briefly appears as a cameo in "A Brave New World"), considering the decent number of viewers (more than thirty thousand) the film secured on its opening day.[7] Box office profits, however, rapidly dropped over the first week of screening, and the film soon disappeared into obscurity.

Given that the South Korean viewership was at least peripherally familiar with the compilation form after successful precedents such as *3 monsŭt'ŏ* (*Three Extremes,* 2004) or *Tokyo!* (2008), *Doomsday Book*'s commercial debacle should be attributed to the stylistic inconsistency of the narrative threads rather than the film's relatively unconventional structure in the context of mainstream South Korean cinema. The marketing tactic also clashed with the overall mood of the film; instead of the collage of diverse perspectives, styles, and tones suggested by the original Korean title *Illyu myŏlmang pogosŏ* (a more faithful translation of which would be "A report on the destruction of humanity"), the film's posters characterized *Doomsday Book* comprehensively as a blockbuster-like epic.[8] Spectators who came to the theaters expecting to see a local rendition of the intricate world-building operations in *The Walking Dead* (2010–present) or a sneak peek at the sublime rush of *World War Z* (2013), the mind-bogging CGI display in *I, Robot,* or the cosmic scale of *Deep Impact* (1998) and *Armageddon* (1998) unfolding on the familiar streets of Seoul, were disappointed by the kitschy tonality of *Doomsday Book* (particularly the bookending episodes "A Brave New World" and "Happy Birthday"). "The Heavenly Creature" digressed from the other two episodes by opting for a more serious mood and heavier message, which became a cause for distraction and confusion rather than serving as a complement. Moreover, with graphics heavily reminiscent of *I, Robot,* stills from "The Heavenly Creature" promised a grand battle between humanity and its mechanical simulations, but instead viewers were treated to long-winded philosophical elocutions in the stationary setting of a religious institution.[9] Independently, "The Heavenly Creature" achieved critical acclaim and won the Cheval Noir prize at the Fantasia Film Festival in 2012, but with the other two episodes eroding its somberness with exaggerated ebullience, *Doomsday Book* as a whole received mediocre to negative reviews.

Nevertheless, *Doomsday Book* merits further scrutiny despite such shortcomings because it opens up a new way of understanding science and technology as a disruptive yet integral force of life within the larger current of South Korean SF cinema. From cellular memories of physical attraction in "A Brave New World" to the bodily fetters of our earthly existence as the seat of spiritual transcendence in "The Heavenly Creature," and cosmic encounters triggered by utterly mundane e-commerce transactions in "Happy Birthday," the three tales in *Doomsday Book* each portray distinct modes of vitality as a mobile force of interchange that defies intentional design or instrumentalization. This take considerably differs from the greater body of SF cinema in South Korea, the characteristics of which, according to Darko Suvin, can be labeled modern.

Suvin argues that modern SF *cognitively estrange* the reader or viewer from the epistemological system of human science by challenging its operative mechanism, which he elaborates on as "the political, psychological, and anthropological *use and effect of knowledge, of philosophy of science,* and the becoming of failure of new realities as a result of it" (emphasis in the original).[10] The modifier "modern" Suvin appends to "science fiction" is at once a paradigmatic signifier and the marker of the temporal juncture he speaks from (the 1970s). The spirit of modern SF stems, Suvin asserts, from the knowledge of the real and attempts to offer variations or critiques of it; even as they aspire to the fantastic (unreal or beyond real), they ultimately circle back to reality. This tendency has been faithfully adopted by mainstream modern Korean SF cinema, many of which remain caught in what Jean Francois Lyotard called "metanarratives" in his attempt to distinguish the postmodern condition from the modern: "a universal history of spirit, in the context of which 'life' is its own self-presentation and formulation in the ordered knowledge of all of its forms contained in the empirical sciences."[11] In South Korean science fiction films that fall under this category, either science and technology is deployed to either celebrate material progress or impart dire warnings against its destructive potentials, or genre tropes are appropriated for dramatic enhancement, in which case plot devices such as time slip, alternative history, or subjectivized perceptions of reality intensify/necessitate the discord structure to enrich the status quo. In both instances, science provides a universally applicable platform of cognition, implemented through human agency via technology, either disrupting or restoring balance. Modern SF, in other words, identifies areas of arrested development (whether it be problematic assimilations to the Western brand of modernity or its failure) in national governance, culture, and society and attempts to repossess the modern by virtue of a better or worse (value-weighted) past or future, seeking remedies for the contemporary.

The presupposition that reality has been estranged from normality because of its digression from desirable metanarratives (such as the creed of modernization) and therefore requires willful intervention also effectively aligns with the tradition of critical realism in South Korea. Although critical realism is a signature trait of South Korean literature, the impulse to critique via representation has also bled into the film sphere, engendering two types of narrative appropriations within the genre. The first category of modern SF cinema in South Korea adopts this approach, using allegorical instances of rogue scientific ventures. The most prominent example would be Bong Joon-ho's globally acclaimed film *Kwoemul* (*The Host,* 2006), in which a mutant creature born of toxic spillage from a US military base wreaks havoc on the citizens of Seoul.[12] The fictional premise calls for the prefix "speculative" rather than "science," a distinction used by writers such as Harlan Ellison or Margaret Atwood to allow for a broader application of plausibility without being constrained by genre conventions or strict scientific proof or reasoning, in that the technological provenance of the creature is secondary to its symbolic significance as the embodiment of the Korean peninsula's colonial vestiges. Naturally, this particular breed of modern SF often revolves around the structural framework of disaster fiction, in which the darker remnants of modernity—such as shadows of imperialism, prioritization of economic prosperity over democracy, and impenetrable bureaucracy—result in serious calamities. Other recent similar films, such as *Yŏn'gasi* (*Deranged,* 2012), *Kamgi* (*The Flu,* 2013), or

Pusanhaeng (*Train to Busan,* 2016), follow a similar path in their portrayal of dysfunctional governance and corporate practices; Bong Joon-ho's *Snowpiercer* (2013) further expands the purview of critical scrutiny by directing its lens toward neoliberalist corpocracy gone global. In these films, the object of estrangement is reality itself. In each instance, seemingly trivial oversights perpetrated by the hands of powerful institutions devolve into massive disasters, threatening to obliterate the society such systems purportedly serve, ranging from misguided capitalism to environmental pollution, bureaucracy, and national security. The trope of normalcy under threat harkens back to metanarratives that demand to be rectified and respected for reality to regain stability.

Films such as *Siwŏlae* (*Il Mare,* 2000), *Tonggam* (*Ditto,* 2000), *2009 Lost Memories* (2001), *Yesterday* (2002), *Ch'ŏn'gun* (*Heaven's Soldiers,* 2005), *Yŏrhansi* (*AM 11:00,* 2013), and *Sigan i t'alcha* (*Time Renegades,* 2016) belong to the second category, in that they use genre tropes to spice up an otherwise conventional drama. In *Siwŏlae, Tonggam,* and *Sigan i t'alcha,* time slippage or dislocation function as communicative portals for relational dynamics such as love, friendship, and camaraderie. The same device changes the course of history by rewriting the precolonial conflict structure between Japan and Korea in *Ch'ŏn'gun. 2009 Lost Memories* and *Yesterday* invert the point of historical divergence by respectively projecting the colonial and predivision periods on to futurity. Although the alternate history trajectory in *Ch'ŏn'gun, 2009 Lost Memories,* and *Yesterday* harkens back to unresolved legacies of modernization and its woes as the grand narrative of nationalism propels the stories, the alternate history framework is more catalytic than ideational, appropriated for dramatic enhancement without necessarily pushing the spectators to question the status quo of reality. In this type of SF narrative, the object of estrangement is cognition because individual or collective memories and knowledge of the past and present are subject to question. However, instead of destabilizing the validity of cognition, the films frame their dislocation as abnormalities to be addressed and consequently rectified in service of a better future. The modern drive toward metanarratives, in this regard, persists; whether it be moral responsibility (*Yŏrhansi*), romantic love (*Siwŏlae*/ *Tonggam*/ *Sigan i t'alcha*), or nationalism (*Ch'ŏn'gun, 2009 Lost Memories, Yesterday*), transcendent values are to be upheld at all cost.[13]

Alain Badiou asserts that "science resembles a metaphorical disquisition, because it elaborates a judgment about what is on the basis of a global fiction in which we experience the momentous question of the relation between the structure of the world and the reality of the choices that one believes oneself to be making within it, or the freedoms that one imagines to be exercising within it."[14] Fiction, meanwhile, has the liberty to use the world itself as a metaphor of the beliefs we harbor, extending the reach of reality as we know it. Thus SF (science + fiction) carries the burden of being at once a metaphorical disquisition and its filtered representation. Stefan Sharff claims that the medium of representation also matters, for, unlike word symbols, which are connected to reality metaphorically, the visual imagery constituting cinema reaffirms the existence of reality by way of a "non-metaphorical order of things."[15] Whereas written SF has the advantage of appropriating the inherent flexibility of semiotic communication by pushing the reader to exercise their imagination as an ingrained process, SF cinema has the challenge of not only balancing the fiction and the science, but also luring the viewers to suspend disbelief against the very nature of the medium itself, the mechanism of which relies on verisimilitude. SF has never been a popular mode of representation in South Korean literature, but its cinematic execution has been burdened by having to work against the compulsion to maximize the medium's reality effect in conjunction with the critical realist tradition in a nonmetaphoric fashion, and to serve as a metaphoric disquisition of existent or desired knowledge and its instantiation while using the world it tries to imagine anew as a metaphor of the real. South Korean SF cinema, so far, has been meeting this demand by resorting to allegorical portrayals of the actual rather than branching out to alternative imaginaries of scientific paradigms that contend (not align to) their counterparts in our world.

In *Doomsday Book,* on the other hand, alterity becomes a condition to live with rather than to survive and remedy in all three episodes—spontaneously affective, rather than rationally directive. Here, the object of estrangement is normalcy itself, reality in its tow. The paradoxical entities in the three episodes such as affectionate zombies, enlightened robots, and closed systems that open up to unknown universes from both within and without illuminate the corresponsive relations between long-standing dichotomies that underpin the modern drive to encase the world in neatly cognizable systems. In this light, *Doomsday Book* radically departs not only from the modern breed of its generic peers within the national context, but also from modern SF at large à la Darko Suvin, because the kind of estrangement *Doomsday Book* subjects the viewers to with its threat of total annihilation is an affective rather than a cognitive one. Instead of imposing a system of rules, the episodes focus on affective moments and instantiations that render what it means to be and live indefinable due to the constant inroads of other matters. As "a gradient of bodily capacity . . . that rises and falls not only along various rhythms and modalities of encounter but also through the troughs and sieves of sensation and sensibility," in Gregory Seigworth's and Melissa Gregg's words, affective engagements not only portend the end of life as we know it, but also gesture to a new opening that accommodates it in alternative form.[16]

"A Brave New World": Affectionate Zombies, Alternate Beings

The Cartesian strain of dualism dictates that the mind and body are discrete, the former holding sway over the latter as an epistemological imperative. Without apperception, the body is merely an empty shell of a meat bag. The issue at stake is control; deprived of the power to cognize, perceive, and govern one's material link to the external world, one loses claim to recognizable existence and succumbs to subjective erasure. The roots of the zombie image in cinema originate from colonial imprints of the master-slave narrative in the West Indies within the context of control over subjectivity.[17] The contemporary zombie begins with George Romero's *Night of the Living Dead* (1968), in which a failed space probe exposed to cosmic radiation crashes down on Earth to kindle a massive outbreak. More like raving ghouls than mindless robots, driven by biological instinct than exogenous will, Romero's zombies are the prototypical inspiration for the zombie figure in contemporary visual culture, threatening to reduce civilization to its bestial origins in massive hoards.[18] Starting out as a monstrous mutation of the indentured servant under imperial rule and evolving over time to embody the mind-numbing force of consumer culture and even nature's retribution against humanity's inroads on its ecological order, zombies on screen have become the new cipher of choice in our time, their conveniently empty bodies awaiting fulfillment through transplanted significance.

"A Brave New World" chooses to go with the epidemic scenario with a hint of critique aimed at wasteful consumerism and ecological disregard, riding the tide of the contemporary zombie fad. The apocalyptic framework in the episode relies on the existential angst one sees in the visceral physicality of the deranged creatures.[19] Civilization is bound to collapse when left in the hands of such entities, who have no use for elaborate contraptions designed to satisfy variegated needs to entertain the mind, but the more fundamental threat of annihilation is geared toward the very possibility of civilization, for instinct sees no future—it only craves to satiate its immediate needs. The erasure of futurity in its species-bound trajectory demands scale both in cinematography and topical reach and stretches beyond the threshold of allegorical social critique. This is an aspect that distinguishes "A Brave New World" from its South Korean *modern* precursors in the zombie-subgenre such as *Purhandangdŭl* (*The Villains,* 2007), *Iutchip chombi* (*The Neighbor Zombie,* 2009), or *Misŭt'ŏ Chombi* (*Mr. Zombie,* 2010), wherein unencumbered nationalism, social isolation and alienation, and economic disenfranchisement each serve as the symbolic referent of zombified existence. Also, although the epidemic-driven apocalypse angle has already become a cliché, its transplantation on to domestic soil afforded "A Brave New World" with novelty value as the first of its kind and remained unchallenged until the release of *Train to Busan* in 2016.

In his insightful analysis of the historiography and philosophical implications of the zombie image in cinema, James McFarland remarks that "as the paradoxical remnant of its own extinction, the zombie-image recalibrates the very possibility of allegorical investment in the temporal world and calls into question the ultimate horizon of life itself."[20] Indeed, the one-to-one correspondence between a problematic social practice and its fantastical projection, as we see in South Korean-brand creature horror (monster) films, loses hold in the face of a crisis that dismantles what it means to be alive at all. Is life present in a crumbling body devoid of its originary biology? How might one define what being (human) means when no subject remains to perceive, cognize, and impose such hermeneutic indicators?

Although "A Brave New World" does little to take the subversive function of the zombie image to any new cinematographic height, it does append a notable twist by, to borrow McFarland's term, "recalibrating" the mechanism of estrangement; the object of estrangement shifts from human society to humanity (as in the values humans cherish and absolutize, Cartesian subjectivity and normalcy being the key here among many others) itself, and its agent from cognizing humans to nonsentient lifeforms that defy their scientifically designated role of evolutionary inferiority in the planetary hierarchy through their applied technicity. The chain of events leading up to the outbreak mocks the anthropocentric logic of the recycling industry by allowing the reappropriated objects within its grinding gears to strike back and expose humanity's bestial instincts that underpin the absurd facades of what we see as eco-friendly science. An unidentified pathogen riding the body of a rotten apple, nonchalantly thrown into the food waste bin by the protagonist, avenges its wasted capacity to affect the world through its natural potentials, namely, the propagation of its own species, by setting off a devastating disease that snatches the fruit of reason out of the humankind's minds. The vitality of these *things* floods the world with a new mode of being in the guise of shelled-out remnants of the familiar human individual. This force and capacity to change is a vital estrangement of preconceived notions of what constitutes being (human, as cognizant and sentient creatures), consequently reconfiguring what the end of such modes of existence means outside the narrow purview of an anthropocentric world order. The things, and the new things (zombies) will persist in their existence beyond the demise of human civilization. The vibrancy of materiality in "A Brave New World" reads like an uncannily spot-on answer to Jane Bennett's question, as to how "patterns of consumption [would] change if we faced not litter, rubbish, trash, or 'the recycling,' but an accumulating pile of lively and potentially dangerous matter."[21]

The sequence leading up to the restaurant scene, where Sŏg-u and his blind date Chi-ŭn first meet over a dish of grilled meat, is yet another example of the missed encounter between affectively estranged *things* (or beings rendered *things*) that later strike back with their capacity to affect. Powered and packaged for livestock feed, the contaminated apple arrives at a cow shed. The camera never captures the cows in full feature, and instead presents piecemeal bits of their body parts that remain relevant to the plot's progression (such as their heads and mouths gulping down the feed). Having fulfilled their reserve value by being thus fattened up, they are slaughtered with one swift strike to the head over a split second, with the camera panning out to demarcate a sense of distance from their last agonizing moment. The shot then jumps to hunks of meat hanging on hooks, their surface bearing crisscrossing red marks to denote further compartmentalization for preferred consumption. The camera zooms in and lingers on the opaque whiteness of what had once been the living animal's eyes before cutting to close-ups of beef dishes under preparation, portending how they would eventually reduce the negligent beholder to their own rank: mere chunks of meat.

Sizzling on the grill in front of Sŏg-u and Chi-ŭn, the meat slices exude a mouth-watering allure of savory sustenance, only to be reduced down to a metaphor of carnal desire as Sŏg-u swallows his rising appetite for Chi-ŭn's shy breasts by hurriedly gulping down the pieces. The camera angle and sound effects accentuate the immediacy of the feast as the two engage in a comical competition to secure more slices for their consumption, the screen moving from fragmented shots of their gulping throats, grasping hands, and smacking lips until Sŏg-u finds a shard of apple peel lodged in a piece of cow liver. This is the moment when the "lively and potentially dangerous matter" of recycled things finally unconceal themselves in plain view, but it completely elides attention as the couple continues to feast on the remaining pile of meat and successfully transition the acoustic festivity to a throaty kiss in the following scene. The fractured cuts of objects that become entangled in the chain of contagion push us to question, as Bennett does, "what difference would it make to public health if eating was understood as an encounter between various and variegated bodies, some of them mine, most of them not, and none of which always gets the upper hand?"[22] As banal as the answer may be (epidemics could be prevented!), "A Brave New World" posits that such affective recalibration could go as far as preventing an apocalypse.

Speaking of banality, the lack of grandiosity is another trait that differentiates "A Brave New World" from the epic strain of zombie epidemics. The utter banality of a molding apple sitting in a fridge in a middle-class household stands in stark contrast to any metanarrative gone awry on a social scale as often seen in other contemporary zombie films, such as failed government cover-up conspiracies in *World War Z,* corporate misconduct and irresponsibility in *Train to Busan,* unforeseen consequences of the Cold War–driven space race in *Night of the Living Dead,* or the overall incompetency of bureaucratic control systems in *The Walking Dead.* Banality is a reflexive mechanism that instigates the end of its own reign in "A Brave New World." Shots of apartment recycling complexes, down-to-earth grill houses, and community parks ground the story in utter ordinariness, each of which harbor a sense of happening that awaits instantiation through their transition from one to the next. Suggesting that the "ordinary is a shifting assemblage of practices and practical knowledges, a scene of both liveliness and exhaustion, a dream of escape or of the simple life," Kathleen Steward suggests that "the notion of a totalized system, of which everything is always already somehow a part, is not helpful (to say the least) in the effort to approach a weighted and reeling present."[23] The urgency of a pandemic most certainly qualifies as an instance of such palpitating presence, but so do the most banal affects that enable its occupants to feel and move through the experience, such as the wrenching disgust Sŏg-u feels as he puts out the trash, or Sŏg-u and Chi-ŭn's mutual affection. In fact, what Steward calls "ordinary affects" are the forces that propel the larger arc of the narrative forward, forcing Sŏg-u's hand to dispose of the molding apple and then reuniting him with Chi-ŭn against all odds on the zombie-ridden streets of Seoul.

Positing a world where estranged things affect the estranger, and the normalcy of ordinary affects harbor most deadly consequences, "A Brave New World" closes with Sŏg-u and Chi-ŭn blankly staring at each other amid the raging madness unfolding on the chaotic streets. Cutting from Chi-ŭn's blacked-out eyes, the glassy sheen of which gestures back to the whites of the dead cows hanging from meat hooks, the camera moves up and away and fades out in a bird's-eye view, leaving a verse from the scripture that points to the provenance of its apocalyptic journey: "And the Lord God commanded the man, You are free to eat from any tree in the garden; but you must not eat from the tree of the knowledge of good and evil, for when you eat from it you will certainly die" (Genesis 2:15–16). The postscript's reference to the apple and its symbolism can be misguiding at first glance, considering how the chain reaction of contagion began with ignorance rather than a manifest desire to attain any transcendent knowledge. However, should we read "knowledge" on a broader stroke, as the metanarrative of science in its systemized integration into the social structure, such as the recycling (resource appropriation) industry and consumer economy culture in which everything is reduced down to compartmentalized exchange value, the postscript could be seen as a direct challenge to the modern drive toward the rational world order of normalcy.

When the infestation reaches a tipping point, the anchor of a news program turns to the viewers and muses, holding a Bible on one hand and a chunk of red meat on the other: "The Bible and the meat, one could say these two items each represent sustenance to the body and the soul. When a fissure erupts between the two . . ." (27:13–27:23). Whereas the zombie virus is a perfect tool to instantiate this material-immaterial division, "A Brave New World" enlists ordinary affects to subvert the metanarrative of anthropocentric epistemology in Cartesian dualism, and going a step further, its ontological legacy of anthropocentric world order. The reality-creating mechanism of technology no longer serves scientific principles, but instead reflects the affective disposition of its executor, as showcased in a news report about the characters in online games going inexplicably amok and behaving like zombies themselves in the film. Even the dichotomy between good and evil loses its hold and becomes a value-neutral matter of survival when the society goes down the drain along with its capacity to distinguish between the two, or rather, appreciate their values. And although all seems to be lost, Sŏg-u and Chi-ŭn's reunion at the denouement suggests that what remains of humanity leaves a crack of opening for a new beginning, albeit in an altered form, on the threshold of apocalypse.

In Closing

In *We Have Never Been Modern,* questioning what it means to be "modern," Latour suggests that critical discourses (by modern humanists) that superimpose science on society while trying to ascertain the irreducible (and therefore inevitably elusive) sovereignty of human subjectivity are destined to failure. According to Latour, they are "reductionist because they seek to attribute action to a small number of powers, leaving the rest of the world with nothing but simple mute forces."[24] *Doomsday Book* is a welcome addition to the cultural imaginary within the South Korean context because it branches out beyond the bias toward critical realism or the lowbrow-highbrow divide in cultural artifacts with its quirky treatment of human subjectivity as both powerless in the face of its revolting creations, and persevering, via mutation. Critical realism stems from the notion that reality must be remedied to reinstate a more desirable past or allow for a better future, based on the presupposition that a set category of values to be abided by exists. South Korean literature and media have largely adhered to this perspective over the past century, struggling to detox the nation and its constituents from the legacy of colonization and the ensuing drive to repossess modernity through economic progress and political democratization. Although noble, the cause has ironically reproduced the principles it endeavors to surmount: the creed of anthropocentric modernity itself, which is heavily indebted to the Cartesian subject and its roots in European Enlightenment. *Doomsday Book* can be read as an attempt to break free of this engrained compliance to existing axiology and embrace the possibility of alternative modes of being within the network of bodies that populate our world as they move and mutate to create new plains of reality. Modern science enforces a system on reality, always trailing behind the scope of perceivable presence, whereas the technicity of matter, unfettered, could open doors to new realities, at times spiraling out to uncharted and unexpected territories.

Each of the three short films in *Doomsday Book* defamiliarize the commonly accepted binaries attached to modern science and technology. "A Brave New World" subverts the axiom that science begets a rational ordering of the world with technology as its conveyor by setting off a (literally) mind-shattering disease that, ironically, originates from the symbolic figure of an apple, the forbidden fruit of reason. The role technology serves in the film is of particular import, because of it actively and agentially enacts these ideas instead of deteriorating into eye-catching trappings. Korean popular culture critic Djuna also emphasizes that South Korean SF films should "relinquish their obsessions with action and large capital" in order to survive and develop their unique appeal. "The Heavenly Creature" recuperates physicality against long-standing notions of dualist metaphysics, featuring the brainchild of modern science, namely, an artificially intelligent robot, who repudiates the anthropocentric ontology of instrumental reason through spiritual awakening. "Happy Birthday," with its comedic overtone, pokes fun at the pervading sense of technological disenchantment in our time by hyperbolically equating the opacity with which our trusty cyberspace infrastructure facilitates encounters that defy human intentionality or comprehension. The fancy trappings of civilization crumble in the face of such absurdities, exposing the raw innards of forces that refuse to be tamed or even understood.

In this chapter, I examine how *Doomsday Book* maps out this alternate topography by focusing on "A Brave New World." The episode has been largely underappreciated by critics in contrast to the acclaim "The Heavenly Creature" garnered, and is often grouped with "Happy Birthday" due to their stylistic affinity, but I believe that "A Brave New World" deserves more appreciation for bringing a refreshing new take at the zombie trope without reducing it to a social allegory. *Doomsday Book* embraces the chaotic jumble of nonhuman agencies as alternative modes of vitality with tropes that defy existing frames of intra- and extraspecies dynamics of the human kind, such as ecological rebounds, emergent independent substrates that set a new precedent in evolutionary biology, and a friendly yet clueless extraplanetary life form.

I use the word "frame" in reference to Martin Heidegger's point that modern science "enframes" the world as a way of subjugating our surroundings and their workings to our needs and wishes. In "The Question Concerning Technology," he admits that technology can be seen as a tool, but contrary to the general notion that it is no more than an instrumental enactment of scientific principles, he claims that technology assists in the process of unconcealing being without boxing it into predisposed agendas. On the other hand, Heidegger muses, modern technology enacts the mechanism of modern science: "[mo]dern science's way of representing pursues and *entraps* nature as a calculable coherence of forces" (emphasis added).[25] His view on science remains consistent throughout his oeuvre, for in "Science and Reflection," he again warns how modern science could limit our understanding of reality and being, because it "objectifies reality" and secures "everything in that objectness which is thus capable of being followed out" in sequences.[26] His insights are useful for mapping out the relationship between affect and technology in *Doomsday Book.* As that which comes after sensation, perception, and memory, Brian Massumi interprets affect in light of the body's being one with its transitions rather than a fixed entity. Delving into Spinoza's pioneering studies on affect, Brian Massumi explains affect as "a *power* (or potential) to affect or be affected."[27] Massumi and Clough's interpretation uniformly underscores the interactive nature of physical being, destabilizing the core tenet of modern science in its attempt to crystalize materiality and its instantiations into fixed and thereby interpretable and predictable states. Given its capacity to create, then, technology is more a force that etches such potentials of fluidity into the human condition rather than a blind enforcer of scientific world order. Technology, in this regard, stands with the affective dynamic of change as much as it correlates to the neat cosmos of modern science.

Embodied, technology in *Doomsday Book* defies the reign of reason and thereby harbingers the expiration of the "Great Divide" Bruno Latour refers to in his project to expose the self-annulling paradox of modernity.[28] The division between nature and nonhumans and between culture and humans no longer holds up in a time when technologized expressions of human agency collapse these boundaries. The mindless body of a zombie aspiring to the most refined qualities of human interaction such as affection, a robot's inorganic and therefore lifeless mind begetting sublime spiritual transcendence of what it lacks by achieving the highest forms of mercy, and a closed virtual network (an internet-based e-commerce site) encasing a portal to a world that rests far beyond the system (the terrestrial) it springs from; these self-contradictory systems in the film, to borrow Heidegger's words, unconceal a paradigmatic shift from predestined design to interactive dynamics, showing how affect engenders emergent capacities.

Notes

1 Under colonization, Korean literature did not have recourse to a wide range of representative strategies, constantly subject to ruthless censorship and weighed down by the challenges of surviving imperial Japan's attempts to quash the peninsula's cultural and even linguistic tradition. After independence, grand narratives loomed large as the nation arduously ploughed through the chaotic aftermath of liberation, the Korean War, military dictatorship, and the growing pains of unbalanced economic progress. Kwŏn Yŏng-min notes that the most important objective in postliberation Korean literature was "cleansing out the legacies of colonial culture and building a new national (*minjok*) literature." Kwŏn, *Han'guk hyŏndae munhaksa* [History of modern Korean literature] (P'aju: T'aehaksa, 2000), 32, author's translation. Burdened with the task of "setting the spiritual coordinate of literature," there was little room to explore alternative (fantastic or nonrealistic) modes of expression, especially given that "all the issues in political reality were directly connected to the state of the literary circles" (37, 31, author's translation). Even advocates of "pure literature" (emphasis on the autonomy and aesthetic properties of literature) who were resistant to the instrumentalization of literature in service of political propaganda relied on grand narratives such as humanism, enacting the sense of responsibility literature was tasked to carry out as critical representations of and inspirations for reality. This paradigm of prioritizing realistic portrayals of the immediacy of reality over diverse modes of expression persisted throughout the latter half of the century until the 1990s, when inklings of political democratization and economic progress begot a new generation of writers who began to blur the boundaries between popular culture and high literature while training more focus on individual concerns.

2 Although the critical realist impulse of modern Korean literature may be key to explaining this lag, SF critic Ko Chang-wŏn offers a more publishing industry and fandom-based analysis in his book *Segye kwahak sosŏlsa* (World science fiction history). Refuting existing claims that the pressures of the colonial rule and the Korean War may have deterred genre literature, he lists five factors: too few opportunities or too little motivation to synergize interest in science and science fiction, respectively; the media's lack of insight regarding science and technology culture and passive attitude toward cultivating said culture; social prejudice in precluding SF from the genre category of general literature; the lack of Korean translators who could select acclaimed works from abroad and publish them for the mass market; and the failure of visual and multimedia in the genre, such as film, animation, and comics, in contributing to the expansion of the literary market for

science fiction (author's translation). See Ko, *Segye kwahak sosŏlsa* (Seoul: Ch'aeryun, 2008).

3 The apocalypse comes in a variety of shapes across cultures, periods, and media. In Korea, *Chŏngkamnok* (a prophetic book by an unknown author from the late Chosŏn period) and the belief in *Mirŭk* (Maitreya, a future Buddha in Buddhist eschatology) are two of the most widely known versions. Whereas *Chŏngkamnok* foretold a political regime change under the guise of a mystical prophesy, the *Mirŭk* lore was deeply rooted in Buddhist mythology, consoling the pains of secular life with faith in a Buddhist messiah descending to earth to bring forth salvation. For more on the *Mirŭk* mythology, see Chang Chi-hun, *Han'guk kodae Mirŭk sinang yŏngu* [A study of Korea's ancient beliefs in the Maitreya] (Seoul: Jinmoondang, 1997). For more on Chŏngkamnok, see Ch'oe Nakki, "Chŏngkamnok e nat'anan p'ungsu sasang yŏngu" [A study on feng shui described in Chŏngkamnok], in *Ŏmunnonjip,* edited by Chungangŏ munhakhoe, 48 (2011): 309–330.

4 Patricia Ticineto Clough, introduction to *The Affective Turn: Theorizing the Social,* ed. Patricia Ticineto Clough and Jean Halley (Durham, NC: Duke University Press, 2007), 2.

5 I have two more reasons for forgoing detailed analyses on the other two episodes: "Happy Birthday" falls short of the other two episodes in many respects (screenplay and cinematography in particular); I also wrote in detail about "Readymade Bodhisattva," the short story on which "The Heavenly Creature" is based, in another essay. See Haerin Shin, "Can Nonhuman Substrates Dream of Nirvana," *Symposium* 70, no. 3 (2016): 153–162.

6 Clough, introduction, 2.

7 See the *Korean Film Council* database search page on *Doomsday Book.*

8 A cursory Naver or Google Image search shows various versions of the posters. Search keyword: illyu myŏlmang pogosŏ + yŏnghwa.

9 In an interview, director Kim Chi-ŭn admits that he was in fact directly inspired by the robot designs in *I, Robot.* Chu Sŏngch'ŏl, "Hyŏndae kigye munmyŏng ŭi taerip chijŏm e pulgyo ka itta" [There is Buddhism at the opposite side of contemporary machine civilization], *Cine 21,* April 19, 2012, http://www.cine21.com/news/view/?mag_id=69592.

10 Darko Suvin, *Metamorphoses of Science Fiction: On the Poetics and History of a Literary Genre* (New Haven, CT: Yale University Press, 1979).

11 Jean-François Lyotard, *The Postmodern Condition: A Report on Knowledge* (Minneapolis: University of Minnesota Press, 1984), 34.

12 Throughout the chapter, I use the official English title for films that have been either released in the United States or have been widely known to Anglophone viewers, and the original Korean titles for others.

13 As in the case of modern Korean literature, adaptations or stories inspired by (but not explicitly attributed to) existing works abroad abounded during the formative years of South Korean cinema. *T'umyŏng in ŭi ch'oehu* (1960), *Uju koein wang magwi* (1967), *Robŏt'ŭ t'aekwŏn V* (1976), *Pich'ŏn koesu* (1984), *Oegye esŏ on uroemae* (1986), *T'umyŏng in'gan* (1986), and *Paio man* (1988) are some of the examples. Independent films of smaller scale such as *Nabi* (2001), *Pŭrein weibŭ* (2005), *Pulch'ŏnggaek* (2010), and *Iutchip chombi* (2010) ventured into more exploratory territories but failed to capture the public's attention due to their marginal levels of exposure. Attempts to pioneer blockbuster SF disastrously failed, as seen in the case of *Naech'yurŏl sit'i* (2003) and *Sŏngnyang p'ari sonyŏ ŭi chaerim* (2002). Select works such as *Chigu rŭl chik'yŏra* (2003) and *Ssaibogŭjiman kwaench'ana* (2006) move away from the modern SF category by sublating the grand narratives in that they achieved a moderate level of critical recognition, but their impact on the overall cinema-sphere was meager at best.

14 Alain Badiou, *Cinema* (Cambridge: Polity, 2013), 194.

15 Stephan Sharff, *The Elements of Cinema: Toward a Theory of Cinesthetic Impact* (New York: Columbia University Press, 1982), 172.

16 Gregory J. Seigworth and Melissa Gregg, "An Inventory of Shimmers," in *The Affect Theory Reader,* ed. Melissa Gregg and Gregory J. Seigworth (Durham, NC: Duke University Press, 2010), 2.

17 For a detailed historiography of the zombie lore, its colonial roots, major films in the genre, and the emergence of the zombie figure as a popular culture icon in our time, see James F. Thompson, "The Rise of the Zombie in Popular Culture," in *But If a Zombie Apocalypse* Did *Occur: Essays on Medical, Military, Governmental, Ethical, Economic and Other Implications,* ed. Amy L. Thompson and Antonio S. Thompson (Jefferson, NC: McFarland, 2015), 11–25. Thompson asserts that though the contemporary zombie is often traced back to religious rituals and practices transplanted to the West Indies, a range of archetypes across cultures and periods portend the formation of the modern zombie. The biomedical approach to zombification, Thompson asserts, begins with the "mad scientist" trope as is the case with many creature horror narratives: "among the first of the zombie films to shift the casual agent for the zombie from the supernatural to the 'mad' scientist is *Bowery at Midnight* (1942, director Wallace Fox), starring Bela Lugosi as the malevolent criminal mastermind converting his reanimated corpses to an afterlife of crime" (14).

18 Ironically, *Night of the Living Dead* was not planned as a zombie film. Instead, as Roger Luckhurst points out, "the lone cadaverous creature hanging around the cemetery in the first scene is another exotic: a *ghoul*"

(Arabic for soul-hungry spirits that haunt graveyards). The original film does not refer to zombies, and Romero's main inspiration was in fact *The Last Man on Earth* (1964), a film adaptation of Richard Matheson's apocalyptic vampire novel. The sequel *Dawn of the Dead* (1978), however, was widely released as a zombie film abroad: "it was only retrospectively, then, that Night became a zombie film—indeed, one of the single most transformative texts in the entire history of the zombie figure." See Roger Luckhurst, *Zombies: A Cultural History* (London: Reaktion Books, 2015).

19 The title is an overt reference to Aldous Huxley's timeless classic *Brave New World,* the title of which is identical to that of the episode's in Korean ("Mŏtchin sinsegye"). The latter however is modified by the indefinite article "a" in its English version, which corresponds to the film's overall trajectory of portraying alterity as an alternate of rather than a digression from normalcy

20 James McFarland, "Philosophy of the Living Dead: At the Origin of the Zombie-Image," *Cultural Critique* 90 (Spring 2015): 56.

21 Jane Bennett, *Vibrant Matter: A Political Ecology of Things* (Durham, NC: Duke University Press, 2010), vii.

22 Ibid.

23 Kathleen Stewart, *Ordinary Affects* (Durham, NC: Duke University Press, 2007), 5.

24 See Bruno Latour, *We Have Never Been Modern,* trans. Catherine Porter (Cambridge, MA: Harvard University Press, 1993), 138.

25 See Martin Heidegger, *The Question Concerning Technology and Other Essays,* trans. William Lovitt (New York: Harper and Row, 1977), 21.

26 Ibid., 168.

27 Brian Massumi, *Parables for the Virtual: Movement, Affect, Sensation* (Durham, NC: Duke University Press, 2002), 15.

28 Latour, *We Have Never Been Modern,* 12.

Bibliography

Adas, Michael. *Machines as the Measure of Men: Science, Technology, and Ideologies of Western Dominance.* Ithaca, NY: Cornell University Press, 1990.

Albrecht, Gary, ed. *Encyclopedia of Disability*, 5 vols. Thousand Oaks, CA: SAGE, 2005.

Allen, Horace. *Allen ŭi ilgi* [The diary of Horace Allen], edited by Kim Wŏnmo. Seoul: Dankook University Press, 1991.

An, Tŏkkyun. *Sejong sidae ŭi pogŏn wisaeng* [Public health measures during the reign of King Sejong]. Seoul: Sejong taewang kinyŏm saŏphoe, 1985.

Anderson, Warwick. "Asia as Method in Science and Technology Studies." *EASTS* 6, no. 4 (2012): 445–451.

Andrews, Bridie. *The Making of Modern Chinese Medicine, 1850–1960.* Honolulu: University of Hawai'i Press, 2015.

Avison, O. R. "Disease in Korea." *Korean Repository* 4 (March 1897): 90–94.

Badiou, Alain. *Cinema.* Cambridge: Polity Press, 2013.

Baker, Don. "Impotent Numbers: Korean Confucian Reactions to Jesuit Mathematics." *Korean Journal of the History of Science* 34, no. 2 (2012): 227–256.

———. "Jesuit Science Through Korean Eyes." *Journal of Korean Studies* 4 (1983): 207–239.

———. "Seeds of Modernity: Jesuit Natural Philosophy in Confucian Korea." *Pacific Rim Report* 48 (August 2007): 1–16.

Baling, Luiz M., and Raul P. Valdez. "Reflections on International Leprosy Congresses and Other Events in Research, Epidemiology, and Elimination of Leprosy." *International Journal of Leprosy* 62, no. 3 (1994): 412–427.

Baron, Samuel, ed. *Medical Microbiology*, 4th ed. Galveston: University of Texas Medical Branch, Department of Microbiology, 1997.

Benjamin, Walter. *Illuminations: Essays and Reflections.* Edited by Hannah Arendt, translated by Harry Zohn. New York: Schocken Books, 1988.

Bennett, Jane. *Vibrant Matter: A Political Ecology of Things.* Durham, NC: Duke University Press, 2010.

Bray, Francesca. "Science, Technology and Late Imperial History." *Chinese Historical Review* 24, no. 1 (2017): 93–104.

———. "Technics and Civilization in Late Imperial China: An Essay in the Cultural History of Technology." *Osiris* 13 (1998): 11–33.

———. *Technology and Gender: Fabrics of Power in Late Imperial China.* Berkeley: University of California Press, 1997.

Brennen, Bonnie, and Hanno Hardt, eds. *Picturing the Past: Media, History and Photography.* Urbana: University of Illinois Press, 1999.

Brown, Nik. "Hope Against Hype: Accountability in Biopasts, Presents and Futures." *Science Studies* 16, no. 2 (2003): 3–21.

———. "Shifting Tenses—From Regimes of Truth to Regimes of Hope?" *Configurations* 13, no. 3 (2007): 331–355.

Bu, Liping, and Ka-che Yip, eds. *Public Health and National Reconstruction in Post-War Asia: International Influences, Local Transformations.* New York: Routledge, 2015.

Buchanan, Tom. "The Courage of Galileo: Joseph Needham and the 'Germ Warfare' Allegations in the Korean War." *History* 86, no. 24 (2001): 503–522.

Bunkenborg, Mikkel. "Subhealth: Questioning the Quality of Bodies in Contemporary China." *Medical Anthropology* 33, no. 2 (2014): 128–143.

Burgin, Victor, ed. *Thinking Photography.* New York: Palgrave Macmillan, 1982.

Bussmann, Klaus, and Florian Matzner, eds. *eine DATA base.* Berlin: Edition Cantz, 1993.

Callon, Michel. "Some Elements of a Sociology of Translation: Domestication of the Scallops and the Fishermen of St Brieuc Bay." *Sociological Review* 32, no. 1, suppl. (1984): 196–233.

Chakrabarty, Dipesh. *Provincializing Europe: Postcolonial Thought and Historical Difference.* Princeton, NJ: Princeton University Press, 2000.

Chang, Hyewŏn. *Sanhaksŏro ponŭn Chosŏn suhak* [Mathematics in the Chosŏn dynasty as seen in books on calculation techniques]. Seoul: Kyŏngmunsa, 2007.

Ch'en, Kenneth. "Matteo Ricci's Contribution to, and Influence on, Geographical Knowledge in China." *Journal of the American Oriental Society* 59, no. 3 (1939): 325–359.

Cherry, Judith. *Korean Multinationals in Europe.* Surrey: RoutledgeCurzon, 2000.

Chesler, Phyllis. *Women and Madness,* 3rd ed. New York: Four Walls Eight Windows, 1997.

Cho, Se-Hui. *The Dwarf.* Translated by Bruce Fulton and Ju-Chan Fulton. Honolulu: University of Hawai'i Press, 2015.

Ch'oe, Ch'ang-jo. "Chosŏn hugi sirhakchadŭl ŭi p'ungsu sasang." *Han'guk munhwa* 11 (1990): 469–504.

———. *Han'guk ŭi p'ungsu sasang* [Korean fengshui philosophy]. Seoul: Minŭmsa, 1984.

Ch'oe, In-jin. "Haebang ch'ogi ŭi sajin chapchi yŏn'gu- Kukche Podo wa Sajin Munhwa rŭl chungsim ŭro." *Aura* no. 10 (2003): 8–17.

Ch'oe, Nakki. "Chŏngkamnok e nat'anan p'ungsu sasang yŏngu" [A study on feng shui described in Chŏngkamnok]. In *Ŏmunnonjip,* edited by Chungangŏ munhakhoe, 48 (2011): 309–330.

Ch'oe, Yŏl. *Han'guk hyŏndae misul pip'yŏngsa.* Kyŏnggido P'ajusi: Ch'ŏngnyŏnsa, 2012.

———. *Han'guk kŭndae misurŭi yŏksa.* Seoul: Yŏrhwadang, 2006.

Choi, Chang-jo. "Study of How Koreans View and Utilize Nature." *Korea Journal* 32, no. 4 (Winter 1992): 26–45.

Chŏn, Sangun. *Sejong sidae ŭi kwahak* [Science during the time of King Sejong]. Seoul: Sejong taewang kinyŏm saŏphoe, 1986.

Chŏng, Chu-yŏng. "P'asŭk'yura ŭi misullon." *Misul Iron'gwa Hyŏnjang* 5 (December 2007): 43–80.

Chŏng, Ku-sŏng. *Chosŏn ŭi memento mori: Chosŏn i bŏrin chadŭl ŭi chukŭm ŭl kiyŏk hara* [Chosŏn Dynasty: Memento Mori: Remembering the deaths of those who were discarded by the Chosŏn dynasty]. Seoul: Aepŭlbuksu, 2010.

Clough, Patricia Ticineto, and Jean Halley, eds. *The Affective Turn: Theorizing the Social.* Durham, NC: Duke University Press, 2007.

Cochrane, R. G. "Editorial—Sulphone Treatment of Leprosy." *International Journal of Leprosy* 17, no. 3 (1949): 299–304.

Commission of International Association of Democratic Lawyers. "Report on U.S. Crimes in Korea, March 31, 1952." P'yôngyang: International Association of Democratic Lawyers, 1952.

The Committee for the Compilation of the History of Korean Women. *Women of Korea: A History from Ancient Times to 1845.* Translated by Yun-Chung Kim. Seoul: Ewha Womans University Press, 1976.

Cooper, Melinda. *Life as Surplus: Biotechnology and Capitalism in the Neoliberal Era.* Seattle: University of Washington Press, 2008.

Cullather, Nick. *The Hungry World: America's Cold War Battle Against Poverty in Asia.* Cambridge, MA: Harvard University Press, 2010.

Daws, Gavin Daws. *Holy Man: Father Damien of Molokai.* Honolulu: University of Hawai'i Press, 1984.

Dawson, Graham. *Soldier Heroes: British Adventure, Empire and the Imagining of Masculinities.* London: Routledge, 1994.

Decker-Phillips, Edith. *Paik Video.* Translated by Marie-Genviève Iselin, Karin Koppensteiner, and George Quasha. New York: Barrytown, 1998.

DiMoia, John P. "From Minnesota to Seoul? The DeWall Helix Bubble Oxygenator and Technology Transfer in Open-Heart Surgery, 1955–1965." *Comparative Technology Transfer and Society* 7, no. 2 (August 2009): 201–225.

Dumit, Joseph. *Drugs for Life: How Pharmaceutical Companies Define Our Health.* Durham, NC: Duke University Press, 2014.

Durkheim, Emile. *The Division of Labor in Society*. Translated by W. D. Halls. New York: Macmillan, 1985.

Eckert, Carter J., et al. *Korean Old and New: A History*. Seoul: Ilchogak, 1990.

Ekbladh, David. *The Great American Mission: Modernization and the Construction of the American World Order*. Princeton, NJ: Princeton University Press, 2011.

Elder, Robert. *The Information Machine: The United States Information Agency and American Foreign Policy*. Syracuse, NY: Syracuse University Press, 1968.

Elkins, James. *Photography Theory*. New York: Routledge, 2007.

Elman, Benjamin A. *On Their Own Terms: Science in China, 1550–1900*. Cambridge, MA: Harvard University Press, 2005.

Evans, Jessica, ed. *The Camerawork Essays: Context and Meaning in Photography*. London: Rivers Oram Press, 1997.

Faget, Guy Henry, F. A. Johansen, and Hilary Ross. "Sulfanilamide in the Treatment of Leprosy." *Public Health Report* 57, no. 50 (1942): 1892–1899.

Faget, Guy Henry, R. C. Pogge, F. A. Johansen, J. F. Dinan, B. M. Prejean, and C. G. Eccles. "The Promin Treatment of Leprosy—A Progress Report." *Public Health Report* 58, no. 48 (1943): 1729–1741.

Fan, Fa-ti. "The Global Turn in the History of Science." *EASTS* 6, no. 2 (2012): 249–258.

———. "Redrawing the Map: Science in Twentieth Century China." *ISIS* 98, no. 3 (2007): 524–538.

Farley, John. *Brock Chisholm, the World Health Organization, and the Cold War*. Vancouver: University of British Columbia Press, 2008.

Fortun, Mike. *Promising Genomics: Iceland and deCode Genetics in a World of Speculation*. Berkeley: University of California Press, 2008.

Gilles, Harold, and D. R. Millard Jr. *The Principles and Art of Plastic Surgery*, 2 vols. Boston, MA: Little, Brown, 1957.

Goeschel, Christian. *Suicide in Nazi Germany*. Oxford: Oxford University Press, 2009.

Goodman, Alan, Deborah Heath, and Susan Lindee, eds. *Genetic Nature/Culture: Anthropology and Science beyond the Two Culture Divide*. Berkeley: University of California Press, 2003.

Gottweis, Herbert, and Robert Triendl. "South Korean Policy Failure and the Hwang Debacle." *Nature Biotechnology* 24, no. 2 (2006): 141–143.

Grant, Edward. "Aristotelianism and the Longevity of the Medieval World View." *History of Science* 16 (1978): 93–106.

Gregg, Melissa, and Gregory J. Seigworth, eds. *The Affect Theory Reader*. Durham, NC: Duke University Press, 2010.

Guillemin, Jeanne. *Biological Weapons: From the Invention of State-sponsored Programs to Contemporary Bioterrorism*. New York: Columbia University Press, 2005.

Ham, Seong Deuk, and Da Sung Yang. "A Study of the Presidential Leadership for Science and Technology in Korea." *Korean Political Science Review* 46, no. 1 (2012): 141–173.

Han, Gil-soo. "The Rise of Western Medicine and Revival of Traditional Medicine in Korea: A Brief History." *Korean Studies* 21(1997): 96–121.

Han, Ki-hyŏng, and Yi Hye-ryŏng, eds. *Yŏm Sang-sŏp Munjang Chŏnjip,* vol. 1. Seoul: Somyŏng Ch'ulp'an, 2013.

Han, Sang-t'ae. "WHO sajŏltan i pon Han'guk pogŏn saŏp—Maektonaldŭ pogosŏ rûl chungsim ŭro" [The public health projects in Korea as seen by the WHO Mission—the MacDonald Report]. *Pogŏn segye* [Public Health World] (April 1960): 40–42.

Han, Tai-Ryoon. "Physical Medicine and Rehabilitation in South Korea: Past, Present, and Future." *American Academy of Physical Medicine and Rehabilitation* 5, no. 11 (November 2013): 964–969.

Han, Young-woo, Ahn Hwi-joon, and Bae Woo Sung. *The Artistry of Early Korean Cartography.* Larkspur, CA: Tamal Vista Publications, 1999.

Hanhardt, John G. *Nam June Paik: Global Visionary.* London: Giles, 2012.

Harley, J. B., and David Woodward, eds. *The History of Cartography,* vol. 2, book 2. Chicago: University of Chicago Press, 1994.

Harris, Steven. *Jesuit Ideology and Jesuit Science: Scientific Activity in the Society of Jesus, 1540–1773.* PhD diss., University of Wisconsin, Madison, 1988.

Heidegger, Martin. *The Question Concerning Technology and Other Essays.* Translated by William Lovitt. New York: Harper and Row, 1977.

Henry, Todd. "Sanitizing Empire: Articulations of Korean Otherness and the Construction of Early Colonial Seoul, 1905–1919." *Journal of Asian Studies* 64, no. 3 (August 2005): 639–675.

———. *Assimilating Seoul: Japanese Rule and the Politics of Public Space in Colonial Korea, 1910–45.* Berkeley: University of California Press, 2014.

Hirschman, Nancy J., and Beth Linker, eds. *Civil Disabilities: Citizenship, Membership, and Belonging.* Philadelphia: University of Pennsylvania Press, 2015.

Hŏ, Chun. *Kukyŏk Tongŭi pogam* [*Tongŭi pogam* with vernacular translation]. Seoul: P'ungnyŏnsa, 1966.

———. "Part I: Internal Bodily Elements." In *Donguibogam: Treasured Mirror of Eastern Medicine,* translated by Kim Namil, Cha Wung Seok, et al. Seoul: Ministry of Health and Welfare, 2013.

Hobsbawm, Eric, and Terence Ranger, eds. *The Invention of Tradition.* Cambridge: Cambridge University Press, 1983.

Ho-Je, Kang. *Pukhan kwahak kisul hyŏngsŏngsa* [History of science and technology in North Korea]. Seoul: Sŏnin, 2007.

Holt, Bertha. *The Seed from the East.* As told to David Wisner. Portland, OR: Holt International Children's Services, 1956.

Hong, Chŏng-sŏn, ed. *Kim P'al-bong Munhak Chŏnjip.* Seoul: Munhak kwa Chisŏngsa, 1989.

Hong, Sungook. "The Relationship between Science and Technology in Korea from the 1960s to the Present Day: A Historical and Reflective Perspective." *EASTS* 6, no. 2 (2012): 259–265.

Hurm, Gerd, Anke Reitz, and Shamoon Zamir. *The Family of Man Revisited: Photography in a Global Age.* London: I. B. Tauris, 2018.

Husserl, Edmund. *On the Phenomenology of the Consciousness of Internal Time (1893–1917).* Translated by John Barnett Brough. Dordrecht: Kluwer Academic Publishers.

Hwang, Kyung Moon. *Beyond Birth: Social Status in the Emergence of Modern Korea.* Cambridge, MA: Harvard University Asia Center, 2004.

———. *Rationalizing Korea: The Rise of the Modern State, 1894–1945.* Berkeley: University of California Press, 2015.

Hyun, Jaewhan. "The Making of a Genetic Nation: Human Heredity, National Identity, and Transnational Scientific Exchange in South Korea, 1926–2009." PhD diss., Seoul National University, 2018.

Im, Chongt'ae. "Chosŏn hugi kwahaksa yŏngu ŭi chŏnjaeng kwa kwaje" [Issues in the history of science in the late Chosŏn period]. *Yŏksa hakpo* 191 (2006): 449–463.

Im, Yŏng-gyun. *Sajin'ga wa ŭi taehwa: Han'guk sajin ŭl kaech'ŏk han wŏllo sajin'ga 8in kwaŭi taedam* 3. Seoul: Nunbit, 1998.

International Scientific Commission for the Investigation of the Facts Concerning Bacterial Warfare in Korea and China. *Report of the International Scientific Commission for the Investigation of the Facts Concerning Bacterial Warfare in Korea and China.* Peking, 1952.

Jasanoff, Sheila, and Sang-Hyun Kim. "Containing the Atom: Sociotechnical Imageries and Nuclear Regulation in the U.S. and South Korea." *Minerva* 47, no. 2 (2009): 119–146.

Jasanoff, Sheila, and Sang-Hyun Kim, eds. *Dreamscapes of Modernity: Sociotechnical Imaginaries and the Fabrication of Power.* Chicago: University of Chicago Press, 2015.

Joen, Sang-woon. "Memories of the Early Days of the Korean History of Science Society, 1960–1980." *Hanguk kwahaksa hoeji* 33, no. 2 (2011): 349–358.

———. *Science and Technology in Korea: Traditional Instruments and Techniques.* Cambridge, MA: MIT Press, 1974.

Jouhki, Jukka Pekka. "Korean Communication and Mass Media Research: Negotiating the West's Influence." *International Journal of Communication,* 2, no. 1 (2008): 253–275.

Kang, Ho-Je. *Pukhan kwahak kisul hyŏngsŏngsa* [History of science and technology in North Korea]. Seoul: Sŏnin, 2007.

Keller, Ulrich F. "The Myth of Art Photography: A Sociological Analysis." *History of Photography* 8, no. 4 (October-December 1984): 249–275.

Kiernan, Ben, ed. *Burchett: Reporting the Other Side of the World, 1939–1983.* London: Quartet Books, 1986.

Kim, Byung-Kook, and Ezra F. Vogel, eds. *The Park Chung Hee Era: The Transformation of South Korea.* Cambridge, MA: Harvard University Press, 2011.

Kim, Dong-no, John B. Duncan, and Kim Do-hyung, eds. *Reform and Modernity in the Taehan Empire.* Seoul: Jimoondang, 2006.

Kim, Dong-won. "The Conflict Between the Image and Role of Physics in South Korea." *Historical Studies in the Physical and Biological Sciences* 33, no. 1 (2002): 107–129.

Kim, Eleanna. *Adopted Territory: Transnational Korean Adoptees and the Politics of Belonging.* Durham, NC: Duke University Press, 2010.

Kim, Eunjung. *Curative Violence: Rehabilitating Disability, Gender, and Sexuality in Modern Korea.* Durham, NC: Duke University Press, 2017.

Kim, Geun Bae Kim. "An Anatomical Chart of South Korean Science and Technology in the 1960s." *EASTS* 5, no. 4 (2011): 529–542.

Kim, Ho. "1612 nyŏn onyŏk palsaeng kwa Ho Chun ŭi Sinch'an pyŏgonbang" [Joseon's response to the 1612 plague—examination of Heo Jun's Shinchan Byeong-onbang]. *Chosŏn sidae sahakpo* 74 (2015): 307–333.

———. "18 segi huban kŏgyŏng sajok wisaeng kwa ŭiryo." *Sŏŭl hak yŏngu* 11 (1998): 113–144.

———. "Chŏngjo tae ŭiryo chŏngch'aek." *Hanguk hakpo* 22, no. 1 (1996): 1231–1257.

———. *Hŏ Chun ŭi Tongŭi pogam yŏn'gu* [A study of Hŏ Chun's Tongŭi pogam]. Seoul: Ilchisa, 2000.

Kim, Hoi-Eun. *Doctors of Empire: Medical and Cultural Encounters between Imperial Germany and Meiji Japan.* Toronto: University of Toronto Press, 2014.

Kim, Jane S.H. "Leprosy in Korea: A Global History." PhD diss., University of California, Los Angeles, 2012.

———. "Desegregate at All Cost!—The World Health Organization and the Battle for Leprosy Control in South Korea, 1966–1982." Paper presented at the Conference of the Association of Asian Studies in Asia, Taipei, June 23, 2015.

Kim, Jea-won. "Riŏllijŭm misul." *Journal of Korean Modern and Contemporary Art History* 22 (December 2011): 71–89.

Kim, Ki-rim. "Han'gang indogyo" [Han River footbridge]. In *Kim Ki-rim Chŏnjip* 1. Seoul: Namsŏldang, 1988.

Kim, Kŭnbae [Geun Bae Kim]. "Pak Chŏnghŭi chŏngbu sigi kwahak kisulŭl ŏttŏk'e pol kŏs'inga? Kwahak taet'ongryŏng tamnonŭl nŏmŏsŏ" [How to understand science and technology of the Park Chung Hee Regime: Beyond the discourse of the 'president of science']. *Yŏksa pip'yŏng* (Spring 2017): 143.

Kim, Leo. "Explaining the Hwang Scandal: National Scientific Culture and Its Global Relevance." *Science as Culture* 17, no. 4 (2008): 397–415.

Kim, Min-chŏng. "Riŏllijŭm ŭi kangbak, chŭngsang ŭrosŏ ŭi riŏllit'i." *Minjok munhaksa yŏn'gu* 54 (2014): 315.

Kim, Okkyun. "Ch'ido yangnon." Reprinted in *Han'guk ŭi kŭndae sasang: Han'guk sasang chŏnjip* 6. Seoul: Samsŏng ch'ulp'ansa, 1984.

Kim, Sang-Hyun. "Pak Chŏnghŭi chŏnggwŏn sigi chŏhang seryŏk ŭi sahoe kisulchŏk sangsang" [Sociotechnical imaginaries and dissident groups during the Park Chung Hee regime], *Yŏksa bip'yŏng* (Fall 2017): 316–346.

Kim, Sonja M. *Imperatives of Care: Women and Medicine in Colonial Korea.* Honolulu: University of Hawai'i Press, 2019.

———. "Missionaries and 'A Better Baby Movement' in Colonial Korea." In *Divine Domesticities: Christian Paradoxes in Asia and the Pacific,* edited by Margaret Jolly and Hyaeweol Choi. Canberra: Australian National University Press, 2014.

Kim, Sung-kyun. *Winding River Village: Poetics of A Korean Landscape.* Privately published in Seoul, 1988.

Kim, Tae-Ho. "How Could a Scientist Become a National Celebrity? Nationalism and Hwang Woo-Suk Scandal." *East Asian Science, Technology and Society,* 2, no. 1 (2008): 27–45.

Kim, Tujong. *Han'guk ŭihaksa* [The history of medicine in Korea]. Seoul: T'amgudang, 1981.

Kim, Yangsu. "Chosŏn hugi chŏnmunjik chungin ŭi Cholla chibanggwan chinch'ul." *Yŏksa wa sirhak* 17–18 (2000): 369–412.

Kim, Yong Woon. "Pan-Paradigm and Korean Mathematics in the Chosŏn Dynasty." *Korea Journal* 26, no. 3 (March 1986): 25–46.

Kim, Yongun, and Kim Yongguk. *Han'guk suhaksa.* Seoul: Yŏlhwadang, 1982.

Kim, Yung Sik. *The Natural Philosophy of Chu Hsi (1130–1200).* Philadelphia, PA: American Philosophical Society, 2000.

———. "Problems and Possibilities in the Study of the History of Korean Science." *Osiris* 13 (1998): 48–79.

———. *Questioning Science in East Asian Contexts: Essays on Science, Confucianism, and the Comparative History of Science.* Leiden: Brill, 2014.

———. "Science and the Confucian Tradition in the Work of Chŏng Yagyong." *Tasanhak* 5 (2004): 127–166.

———. "Specialized Knowledge in Traditional East Asian Contexts." *EASTS* 4, no. 2 (2010): 179–183.

———. "The 'Why Not' Question of Chinese Science: Scientific Revolution and Traditional Chinese Science." *East Asian Science, Technology and Medicine* 22 (2004): 96–112.

Kitanaka, Junko. *Depression in Japan: Psychiatric Cures for a Society in Distress.* Princeton, NJ: Princeton University Press, 2012.

Kleinman, Daniel Lee. *Science and Technology in Science: from Biotechnology to the Internet.* Victoria: Blackwell Publishing, 2005.

Kleinman, Daniel Lee, and Kelly Moore, eds. *Routledge Handbook of Science, Technology and Society.* London: Routledge, 2014.

Koo, Hagen, ed. *State and Society in Contemporary Korea.* Ithaca, NY: Cornell University Press, 1994.

Korea Child Welfare Committee, Han Welfare Committee. *Handicapped Children's Survey Report, Korea, 1961* [Han'guk changhae adong chosa pogosŏ]. Seoul: Pogŏn Sahoebu Han'guk Adong Pongni Wiwŏnhoe, 1961.

Kuksa p'yŏnch'an wiwŏnhoe, ed. *Han'guksa* [The history of Korea], vol. 10. Seoul: T'amgudang, 1990.

Kushner, Howard. "Suicide, Gender and the Fear of Modernity in Nineteenth-Century Medical and Social Thought." *Journal of Social History* 6, no. 3 (1993): 461–490.

Kwŏn, Yŏng-min. *Han'guk hyŏndae munhaksa* [History of modern Korean literature]. P'aju: T'aehaksa, 2000.

Lacina, Bethany, and Nils P. Gleditsch. "A New Dataset of Battle Deaths." *European Journal of Population* 21 (2005): 154.

Lada, John, ed. *Civil Affairs/Military Government Public Health Activities.* Preventive Medicine in World War II, vol. 8. Washington, DC: Office of the Surgeon General, Department of the Army, 1976.

Latour, Bruno. *We Have Never Been Modern.* Translated by Catherine Porter. Cambridge, MA: Harvard University Press, 1993.

Lee, Young Hee. "How Much Korea-Specific Phenomenon Is Hwang's Scandal? Reading Universality and Specificity of Hwang's Scandal." *Journal of Science & Technology Studies* 7, no. 2 (2007): 31–42.

Leeke, Ada. *When Americans Came to Korea.* Freeman, SD: Pine Hill Press, 1991.

Leitenberg, Milton. "New Russian Evidence on the Korean War Biological Warfare Allegations: Background and Analysis." *Cold War International History Project Bulletin* 11 (1998): 185–197.

Levin, Thomas Y., ed. *The Mass Ornament: Weimar Essays.* Cambridge, MA: Harvard University Press, 1995.

Li, Yan, and Du Shiran. *Chinese Mathematics: A Concise History.* Translated by John N. Crossley and Anthony W.C. Lun. Oxford: Clarendon Press, 1987.

Lieberman, Lisa. *Leaving You: The Cultural Meaning of Suicide.* New York: Ivan R. Dee, 2003.

Limb, Eung-Sik. *Limb Eung-Sik (Im Ŭng-sik) hoegorok: Naega kŏrŏ on han'guk sadan.* Seoul: Nunbit, 1997.

Linker, Beth. *War's Waste: Rehabilitation in World War I America.* Chicago: University of Chicago Press, 2011.

Lockwood, Diana N.J. "Human Immunodeficiency Virus and Leprosy: An Update." *Dermatologic Clinics* 29, no.1 (2011): 125–128.

Luckhurst, Roger. *Zombies: A Cultural History.* London: Reaktion Books, 2015.

Lyotard, Jean-François. *The Postmodern Condition: A Report on Knowledge.* Minneapolis: University of Minnesota Press, 1984.

Marsh, Ian. *Suicide: Foucault, History and Truth.* Cambridge: Cambridge University Press, 2010.

Massumi, Brian. *Parables for the Virtual: Movement, Affect, Sensation.* Durham, NC: Duke University Press, 2002.

McFarland, James. "Philosophy of the Living Dead: At the Origin of the Zombie-Image. *Cultural Critique* 90 (Spring 2015): 22–63.

Merle, Corinee S.C., Sergio S. Cunha, and Laura C. Rodrigues. "BCG Vaccination and Leprosy Protection: Review of Current Evidence and Status of BCG in Leprosy Control." *Expert Review of Vaccines* 9, no. 2 (2010): 209–222.

Millard, D. Ralph. "Oriental Peregrinations." *Plastic and Reconstructive Surgery* 16, no. 5 (November 1955).

Millett, Allen. *The War for Korea, 1945–1950: A House Burning.* Lawrence: University Press of Kansas, 2005.

Min, Sŏng-kil. *Hwabyŏng yŏn'gu* [Study of fire illness]. Seoul: ML, 2009.

Mizuno, Hiromi. "Introduction." *EASTS* 6, no. 1 (2012): 1–8.

———. *Science for the Empire: Scientific Nationalism in Modern Japan.* Palo Alto, CA: Stanford University Press, 2008.

Mizuno, Hiromi, Aaron S. Moore, and John DiMoia, eds. *Engineering Asia: Technology, Colonial Development and the Cold War Order.* London: Bloomsbury Publishing, 2018.

Moon, Manyŏng. "Understanding Compressed Growth of Science and Technology in South Korea: Focusing on Public Research Institutes." *Hanguk kwahak sahak hoeji* 37, no. 2 (2015): 431–453.

Moon, Manyŏng and Kim Yŏngsik. *Han'guk kŭndae kwahak hyŏngsŏng kwajŏng charyo* [Documentary history of the development of modern science in Korea]. Seoul: Seoul National University Press, 2004.

Nakayama, Shigeru. *A History of Japanese Astronomy: Chinese Background and Western Impact.* Cambridge, MA: Harvard University Press, 1969.

National Human Rights Commission. *Hansenin inkwôn silt'ae chosa* [Investigation on the conditions of human rights of the Hansen's disease patients]. Seoul: National Human Rights Council of Republic of Korea, 2005.

———. "Hansenin kwŏnik kanghwa mit chngch'akch'on hwan'gyŏng kaesŏn pang'an" [Increase of the rights of the Hansen's disease patients and improvement of resettlement village environment]. Seoul: Bureau of Welfare and Labour Complaints, Anti-Corruption & Civil Rights Commission, June 2010.

Needham, Joseph, Lu Gwei-Djen, John Combridge, and John Major. *The Hall of Heavenly Records: Korean Astronomical Instruments and Clocks, 1380–1780.* Cambridge: Cambridge University Press, 1986.

Needham, Joseph. *Science and Civilization in China,* vol. 3. London: Cambridge University Press, 1959.

Oh, Arrisa. *To Save the Children of Korea: The Cold War Origins of International Adoption.* Stanford, CA: Stanford University Press, 2015.

Oh, Hye-ri. "The Concept of Photography in Korea: The Genealogy of Korean Conception of the Korean Conception of *Sajin* from the late Chosŏn Dynastic Period Through Japanese Colonialism." PhD diss., Binghamton University, State University of New York, 2014.

———. "Translating 'Photography': The Migration of the Concept of *Sajin* from Portraiture to Photography." *History of Photography* 39, no. 4 (November 2015): 366–389.

Oh, Sang-Hak. "Circular World Maps of the Joseon Dynasty: Their Characteristics and Worldview." *Korea Journal* 48, no. 1 (Spring 2008): 8–45.

———. "The Recognition of Geomancy by Intellectuals during the Joseon Period," *Review of Korean Studies* 13, no. 1 (March 2010): 121–147.

Paek, Unsok. "The Empiricist's Progress." *Yugyo munhwa yŏn'gu* [Journal of Confucian philosophy and culture] 8 (2007): 231–260.

Paik, Nam June. *Videa 'n' Videology 1959–1973,* edited by Judson Rosebush. Syracuse, NY: Everson Museum of Art, 1974.

Pak, Chong-sŏng. *Sarang hada chukda: Chŏngsa ŭi chongch'ihak* [To love and then to die: The politics of love suicides]. Seoul: In'gan sarang, 2012.

Pak, P'il-ho. *Sajin ŭl mal handa.* Seoul: Sigak, 2003.

Pak, T'aewŏn. *Sosŏlga Kubo-ssi ŭi iril* [A day in the life of Kubo the novelist]. Seoul: Muhak kwa chisŏngsa, 2005.

Pak, Yunjae. *Hanguk kŭndae ŭihak ŭi kiwon* [The origin of the Korean modern medical system]. Seoul: Hyean, 2005.

———. "Yangsaengesŏ wisaengŭro, kaehwap'a ŭi ŭihangnon kwa kŭndae kukka kŏnsŏl" [From *yangsaeng* to *wisaeng,* medical discourse of the Kaehwa group and construction of the modern state]. *Sahoe wa yŏksa* 63 (2003): 30–50.

Park, Chang-won. "Funerary Transformations in Contemporary South Korea." *Mortality* 15, no. 1 (February 2010): 18–37.

Park, Jaeyung, Hyoungjoon Jeon, and Robert A. Logan. "The Korean Press and Hwang's Fraud." *Public Understanding of Science* 18, no. 6 (2009): 653–669.

Park, Jin-kyung. *Corporeal Colonialism: Medicine, Reproduction, and Race in Colonial Korea.* PhD diss., University of Illinois at Urbana-Champaign, 2008.

Park, Seong-Rae. "The Introduction of Western Science in Korea, 1876–1910." *Korea Journal* 21, no. 5 (May 1981): 29–38.

———. "Portents in Korean History." *Journal of Social Sciences and Humanities* 47 (June 1978): 31–92.

———. "Portents and Neo-Confucian Politics in Korea, 1392–1519." *Journal of Social Sciences and Humanities* 49 (June 1979): 53–117.

———. *Portents and Politics in Korean History.* Seoul: Jimoondang, 1998.

———. "Rise of Confucian Porentology in Korea." *Journal of Social Sciences and Humanities* 48 (December 1978): 1–45.

Pulfrey, Caroline, and Fabrizio Butera. "Why Neoliberal Values of Self-Enhancement Lead to Cheating in Higher Education: A Motivational Account." *Psychological Science* 24, no. 11 (2013): 2153–2162.

Rajan, Kaushik Sunder. *Biocapital: The Constitution of Postgenomic Life.* Durham, NC: Duke University Press, 2006.

Rees, R.J.W., and A. C. McDougall. "Airborne Infection with *Mycobacterium Leprae in* Mice." *Journal of Medical Microbiology* 10 (1977): 63–68.

Rees, R.J.W., J.M.H. Pearson, and M.F.R. Waters. "Experimental and Clinical Studies on Rifampicine in Treatment of Leprosy." *British Medical Journal* 1, no. 5688 (1970): 89–92.

Rodrigues, Laura C., and Diana N.J. Lockwood. "Leprosy Now: Epidemiology, Progress, Challenges, and Research Gaps." *The Lancet: Infectious Diseases* 11, no. 6 (2011): 464–470.

Roetz, Heiner, ed. *Cross-Cultural Issues in Bioethics: The Example of Human Cloning.* New York: Rodopi B.V., 2006.

Rogaski, Ruth. *Hygienic Modernity: Meanings of Health and Disease in Treaty-Port China.* Berkeley: University of California Press, 2004.

———. "Nature, Annihilation, and Modernity: China's Korean War Germ Warfare Experience Revisited." *Journal of Asian Studies* 61, no. 2 (2002): 381–415.

Rusk, Howard. *A World to Care For.* New York: Random House, 1972.

Sakae, Miki. *Chōsen igakushi oyobi shippeishi* [History of Korean medicine and disease]. Osaka: Shibun chuppansha, 1962.

Sakai, Naoki. "Theory and Asian Humanity: On the Question of *Humanitas* and *Anthropos.*" *Postcolonial Studies* 13, no. 4 (2010): 441–464.

Sandeen, Eric J. *Picturing an Exhibition: The Family of Man and 1950s America.* Albuquerque: University of New Mexico Press, 1955.

Scheid, Volker. *Currents of Tradition in Chinese Medicine: 1626–2006*. Seattle, WA: Eastland Press, 2007.

Schmid, Andre. *Korea Between Empires, 1895–1919*. New York: Columbia University Press, 2002.

Schreuder, Pieter A.M., Salvatore Noto, and Jan Hendrik Richardus. "Epidemiologic Trends of Leprosy for the 21st Century." *Clinics in Dermatology* 33, no. 1 (2016): 24–31.

Scott, James. *Weapons of the Weak: Everyday Forms of Peasant Resistance*. New Haven, CT: Yale University Press, 1985.

Sekula, Allan. "Traffic in Photographs." In *Photography against the Grain: Essays and Photo Works 1973–1983*. Halifax: The Press of the Nova Scotia College of Art and Design, 1984.

Setia, Maninder Singh, Craig Steinmaus, Christine S. Ho, and George W. Rutherford. "The Role of BCG in Prevention of Leprosy: A Meta Analysis." *The Lancet: Infectious Diseases* 6, no. 3 (2006): 162–170.

Sharff, Stefan. *The Elements of Cinema: Toward a Theory of Cinesthetic Impact*. New York: Columbia University Press, 1982.

Shin, Dongwon. "The Characteristics of Joseon Medicine Discourses on the Body, Illustration and Dissection." *Review of Korean Studies* 13, no. 1 (Spring 2010): 149–178.

———. "Chosŏn sidae ŭi ŭihangnon" [Medical theory during the Chosŏn period]. *Ŭisahak* 13, no. 1 (2004): 134–145.

———. "Hanguk chŏnt'ong ŭihak ŭi ŭihak yulli wa saengmyŏng yulli: Chosŏn sidae ŭi ŭisŏrŭl chungsimŭro" [Bioethics and medical ethics in Korean medical tradition]. *Chonggyo munhwa pip'yŏng* 5 (2004): 65–90.

———. *Hanguk kŭndae pogŏn ŭiryosa* [History of health and medicine in modern Korea]. Seoul: Hanul Academy, 1997.

———. *Hoyŏlja Chosŏnŭl sŭpkyŏkhada: mom kwa ŭihak ŭi Hanguksa* [Cholera invades Chosŏn: History of the body and medicine in Korea]. Seoul: Yŏksa pip'yŏngsa, 2004.

———. "Korean Anatomical Charts in the Context of the Asian Medical Tradition." Translated by Yuseok Kim. *Asian Medicine* 5 (2009): 186–207.

Shin, Dongwon, and Hwang Sangik. "Chosŏn malgi (1876–1910) kŭndae pogŏn ŭiryo ch'eje ŭi hyŏngsŏng kwajŏng kwa kŭ ŭimi" [Formation of the modern health and medical system in late Chosŏn], *Ŭisahak* 5, no. 2 (1996): 155–165.

Shin, Dongwon, Kim Namil, and Yŏ Insŏk, *Hangwŏnŭiro ingnŭn* Tongŭi pogam [Reading the Tongŭi pogam in one volume]. Seoul: Tŭllyŏk, 1999.

Shin, Haerin. "Can Nonhuman Substrates Dream of Nirvana." *Symposium* 70, no. 3 (2016): 153–162.

Sivin, Nathan. *Traditional Medicine in Contemporary China*. Ann Arbor: Center for Chinese Studies, University of Michigan, 1987.

———. "Why the Scientific Revolution Did Not Take Place in China—or Didn't It?" *Chinese Science* 5 (1982): 45–66.

Sloan, Norman R., Edwin K. Chung-Hoon, M. E. Godfrey-Horan, and G. H. Hedgcock. "Sulfone Therapy in Leprosy: A Three Year Study." *International Journal of Leprosy* 18 no. 1 (1950): 1–9.

Snyder, Joel. "Picturing Vision." *Critical Inquiry* 6 (Spring 1980): 499–526.

Song, Sang-yong. "Beyond Scientism: Coming of the Ethics of Science." *Hanguk kwahaksa hakhoeji* 35, no. 2 (2013): 389–398.

Steichen, Edward. *The Family of Man*. New York: Museum of Modern Art, 1955.

———. "Photography: Witness and Recorder of Humanity." *Wisconsin Magazine of History* 41, no. 3 (Spring 1958): 159–167.

Stewart, Kathleen. *Ordinary Affects*. Durham, NC: Duke University Press, 2007.

Stiegler Bernard. *Technics and Time, 1: The Fault of Epimetheus*. Translated by Richard Beardsworth and George Collins. Stanford, CA: Stanford University Press, 1998.

———. *Technics and Time, 3: Cinematic Time and the Question of Malaise*. Translated by Stephen Barker. Stanford, CA: Stanford University Press, 2011.

Suh, Soyoung. "Korean Medicine Between the Local and the Universal: 1600–1945." PhD diss., University of California, Los Angeles, 2007.

———. *Naming the Local: Medicine, Language, and Identity in Korea since the Fifteenth Century*. Cambridge, MA: Harvard University Asia Center, 2017.

Sung, Byungchang. "The Education Policy of the Participatory Government." *Education Review* 12 (Summer 2003): 22–24.

Suvin, Darko. *Metamorphoses of Science Fiction: On the Poetics and History of a Literary Genre*. New Haven, CT: Yale University Press, 1979.

Suzuki, Koichi, Takeshi Akama, Akira Kawashima, Aya Yoshihara, Rie R. Yotsu, and Norihisa Ishii. "Current Status of Leprosy: Epidemiology, Basic Science, and Clinical Perspectives." *Journal of Dermatology* 39, no. 2 (2011): 121–129.

Sworakowski, Witold S., ed. *World Communism: A Handbook, 1918–1965*. Stanford, CA: Hoover Institution Press, 1973.

Szarkowski, John, and Shoji Yamagishi, eds. *New Japanese Photography*. New York: Museum of Modern Art, 1974.

Tagg, John. *The Burden of Representation*. Minneapolis: University of Minnesota Press, 1993.

Tanne, Janice Hopkins. "Science Will Tighten Standards after Retracting Stem Cell Papers." *British Medical Journal* 333 (2006): 1189.

Thompson, Amy L., and Antonio S. Thompson. *But If a Zombie Apocalypse* Did *Occur: Essays on Medical, Military, Governmental, Ethical, Economic and Other Implications*. Jefferson, NC: McFarland, 2015.

Thompson, Charis. *Making Parents: The Ontological Choreography of Reproductive Technologies*. Cambridge, MA: MIT Press, 2005.

Tucker, Anne Wilkes, et al., eds. *The History of Japanese Photography*. New Haven, CT: Yale University Press, 2003.

Underwood, Horace. *Korea in War, Revolution and Peace: The Recollections of Horace G. Underwood*, edited by Michael J. Devine. Seoul: Yonsei University Press, 2001.

Unschuld, Paul. *Medical Ethics in Imperial China*. Berkeley: University of California Press, 1979.

US Department of State. "The World Peace Council's 'Peace Assemblies,'" *Foreign Affairs Note* (May 1983): 1–6.

Vartanian, Ivan, Akihiro Hatanaka, and Yutaka Kambayashi, eds. *Setting Sun: Writings by Japanese Photographers*. New York: Aperture, 1992.

Vogel, Hans Ulrich, and Günter Dux, eds. *Concepts of Nature: A Chinese-European Cross-Cultural Perspective*. Leiden: Brill, 2010.

Waldby, Catherine. "Stem Cells, Tissue Cultures and the Production of Biovalue." *Health: An Interdisciplinary Journal for the Social Study of Health, Illness and Medicine* 6, no. 3 (2002): 305–323.

Weathersby, Kathryn. "Deceiving the Deceivers: Moscow, Beijing, Pyongyang, and the Allegations of Bacteriological Weapons Use in Korea." *Cold War International History Project Bulletin* 11 (1998): 176–185.

Wernicke, Günther. "The Communist-Led World Peace Council and the Western Peace Movements: The Fetters of Bipolarity and Some Attempts to Break Them in the Fifties and Early Sixties." *Peace & Change* 23, no. 3 (1998): 265–311.

Winchester, Simon. *The Man Who Loved China: The Fantastic Story of the Eccentric Scientist Who Unlocked the Mysteries of the Middle Kingdom*. New York: Harper Collins, 2008.

Wittner, Lawrence S. *The Struggle against the Bomb: One World or None: A History of the World Nuclear Disarmament Movement through 1953*. Stanford, CA: Stanford University Press, 1993.

Wong, Joseph. *Healthy Democracies: Welfare Politics in Taiwan and South Korea*. Ithaca, NY: Cornell University Press, 2004.

World Health Organization. *Fifty Years of the WHO in the Western Pacific Region*. Manila: WHO, 1998.

———. "Global Leprosy Update, 2013: Reducing Disease Burden." *Weekly Epidemiological Record* 36, no. 89 (September 5, 2014): 389–400.

———. *Polio Eradication in the Western Pacific Region*. Manila: WHO, 2002.

———. *WHO Study Group on Epidemiology of Leprosy in Relation to Control*. Geneva: World Health Organization, 1985.

Yang, Bo-Kyung. "Perceptions of Nature in the Chosŏn Period." *Korea Journal* 37, no. 4 (Winter 1997): 134–155.

Yawalkar, S. H., A. C. Mcdougall, J. Languillon, S. Gosh, S. K. Hajra. D.V.A. Opromolla, and C. J. S. Tonnello. "Once-Monthly Rifampin Plus Daily Dapsone in Initial Treatment of Lepromatous Leprosy." *The Lancet* 319, no. 8283 (1982): 1199–1202.

Yi, Ch'an. "Chosŏn sidae ŭi chidoch'aek." In *Han'guk ŭi chŏnt'ong chiri sasang.* Seoul: Minŭmsa, 1992.

Yi, Ch'ŏl-Ho. "Sin Kyŏnghyangp'a pip'yŏng ŭi nangman chuŭijŏk kiwŏn." *Minjok munhaksa yŏn'gu* 38 (2008): 235.

Yi, Hwa. *P'ungsuran muŏsin'ga* [What is geomancy?]. Seoul: Ihaksa, 2013.

Yi, Kidŏk. *Hanguk kŭndae sŏyang ŭihak kyoyuksa* [The early history of modern medical education in Korea]. Seoul: Ch'ŏngnyŏn ŭisa, 2008.

Yi, Kkonme. "Ilbanin ŭi Hanŭhak insk kwa ŭiyak iyong" [A study on the general public understanding and utilization of traditional Korean medicine in the colonial period]. In *Hanŭihak, singminji rŭl alt'a: singminji sigi Hanŭihak ŭi kŭndaehwa yŏngu* [The modernization of Korean traditional medicine during the colonial period], edited by Yonsei University Institute for History of Medicine, 137–154. Seoul: Ak'anet, 2008.

Yi, Kyŏng-min. "Charyoro pon Han'guk sajinsa: Haebang gonggan ŭi nambukhan sajin'gye ŭi chojik (1)." *Sajin & Munhwa* 1 (February 2010): 5.

———. "Charyoro pon Han'guk sajinsa: Haebang konggan'gwa han'guk chŏnjaenggi ŭi nambukhan sajin'gye ŭi chojik (2)." *Sajin & Munhwa* 1 (July 2010): 5.

Yi, Kyŏngnok. "Chosŏn Chungjong 19–20 nyŏn ŭi chŏnyŏmbyŏng ch'anggwŏl kwa kŭ tae'ŭng" [The epidemic outbreak under Chungjong's reign and its responses]. *Chungang saron* 39 (2014): 5–40.

Yi, Kyugŭn. "Chosŏn sidae ŭiryo kigu wa ŭigwan" [Chosŏn period medical bureaucracy and medical officers]. *Tongbang hakji* 104 (1999): 95–161.

Yi, Manyŏl. *Hanguk kidokkyo ŭiryosa* [A history of Christian medical work in Korea]. Seoul: Acanet, 2003.

Yi Pu-yŏng, "Ilcheha chŏngsingwa chillyo wa kŭ pyŏnchŏn: Chosŏn ch'ongdokpu ŭiwŏn ŭi chŏngsingwa chillyo (1913–1928) rŭl chungsim ŭro" [With special reference to the psychiatric care and its change under the Japanese colonial rule: Clinical activities at the government general hospital, 1913–1928], *Ŭisahak* 3, no. 2 (1994): 153–154.

Yi, Yŏng-a. "1920-nyŏndae sosŏl ŭi 'chasal' hyŏngsanghwa yangsang yŏn'gu" [A study on the figurative methods and meanings of 'suicide' in the 1920s novels]. *Han'guk hyŏndae munhak yŏn'gu* 33 (April 2011): 207–248.

Yi Yŏnghŭi [Young Hee Lee]. "Hwang Usŏk sat'aesŭn ŏlmana Han'gukchŏk in'ga?" [How much Korea-specific phenomenon is Hwang's scandal? Reading universality and specificity of Hwang's scandal]. *Kwahak kisulhak yŏnguk* [Journal of Science & Technology Studies] 7, no. 2 (2007): 31–42.

Yŏ, Insŏk, Yi Hyŏnsuk, Kim Sŏngsu, Sin Kyuhwan, Pak Yunhyŏng, and Pak Yunjae. *Han'guk ŭihaksa* [The history of medicine in Korea]. Seoul: Institute for Research into Medical Policy of the Korean Medical Association, 2012.

Yonsei Taehakkyo Kukhak yŏn'guwŏn, ed. *Han'guk sirhak sasang yŏn'gu* IV [Studies in Korea's practical learning]. Seoul: Hyean, 2005.

Yoon, Hong-Key. *The Culture of Fengshui in Korea: An Exploration of East Asian Geomancy.* Lanham, MD: Lexington Books, 2006.

Yoon, Inshil Cho, trans. *A Place to Live: A New Translation of Yi Chung-Hwan's T'aengniji: The Korean Classic for Choosing Settlements.* Honolulu: University of Hawai'i Press, 2018.

Yu, Sanggyu. "Seburansŭ pyŏngwŏn munje rŭl chungsimŭro" [On the Severance Hospital problem]. *Sin Tonga* 5, no. 4 (1935). Reprinted in Yu Sanggyu and Yu Yongsŏp, *Aeguk chisa T'aehŏ Yu Sanggyu: chŏngi wa yugo* [Patriot T'aehŏ Yu Sanggyuk: His life and works]. Seoul: Hongsadan ch'ulp'anbu, 2007.

Yuh, Leighanne Kimberly. "Moral Education, Modernization Imperatives, and the *People's Elementary Reader* (1895): Accommodation in the Early History of Modern Education in Korea." *Acta Koreana* 18, no. 2 (2015): 327–355.

Zhan, Mei. "A Doctor of the Highest Caliber Treats an Illness before It Happens." *Medical Anthropology* 28, no. 2 (2009): 166–188.

Zur, Dafna, and Christopher P. Hanscom. "Science and Literature in Korea: An Introduction." *Journal of Korean Studies* 23, no. 2 (2018): 213–222.

Contributors

Don Baker first became interested in Korea while teaching English in Kwangju as a Peace Corps volunteer in the early 1970s. He received his PhD in Korean history from the University of Washington in 1983. In 1987, he began teaching Korean history and religion in the Department of Asian Studies at the University of British Columbia, where he still teaches today. His research and publications focus on the cultural history of Korea since the sixteenth century; his special interest is in religion, philosophy, traditional science, and the life and thought of Tasan Chŏng Yagyong. Among his publications are articles analyzing traditional astronomy, medicine, and mathematics in the last two centuries of the Chosŏn dynasty and how they were transformed in the late nineteenth and early twentieth centuries. Because he was fortunate enough to witness the transformation of South Korea from poverty and dictatorship to prosperity and democracy, he is also interested in recent political, economic, and social change and has published on the Kwangju uprising of May 1980. His most recent book is *Catholics and Anti-Catholicism in Chosŏn Korea* (University of Hawai'i Press, 2017).

Steve Choe is an associate professor in the School of Cinema at San Francisco State University. His areas of research and teaching include film and media theory, philosophy, and the cinemas of Germany and Korea. He is the author of *Afterlives: Allegories of Film and Mortality in Early Weimar Germany* (2014), *Sovereign Violence: Ethics and South Korean Cinema in the New Millennium* (2016), and *ReFocus: The Films of Williams Friedkin* (2021). With Mayumo Inoue, he is a coeditor of *Beyond Imperial Aesthetics: Theories of Art and Politics in East Asia* (2019).

John DiMoia is an associate professor of Korean history at Seoul National University. Previously, he was an associate professor of history at the National University of Singapore (NUS). He is the author of *Reconstructing Bodies: Biomedicine, Health, and Nation-Building in South Korea since 1945* (2013). Along with Hiromi Mizuno of the University of Minnesota and Aaron S. Moore of Arizona State University, he is a coeditor of *Engineering Asia: Technology, Colonial Development and the Cold War Order* (2018), which examines inter-Asian technical exchange from the interwar years to the late twentieth century. His current project looks at South Korean developmentalism through its relations with Southeast Asia, beginning with civic and medical outreach projects in Thailand and Vietnam conducted during the war (1964–1973), and tracking these efforts to present-day Korean economic expansion in Vietnam, following normalization in the early 1990s.

Inkyu Kang is an associate professor of journalism at Penn State University (Behrend College). Before joining Penn State, he taught mass media and popular culture at the University of Wisconsin-Madison, where he holds a PhD in communication arts. Kang's major research interests are global media, cultural studies, and the sociology of technology. He has published in several academic journals, coauthoring or coediting anthologies including *Sport in Korea: History, Development, Management* (2018), *K-Pop: The International Rise of the Korean Music Industry* (2015), and *Korean Popular Culture Reader* (2014). He is also an award-winning journalist who has written extensively about social and technological issues. Kang is currently conducting research on the evolution of Russia's performing arts since the 1917 revolution.

Jane S. H. Kim is an independent scholar who received her PhD in Korean history (Korean Studies) at University of California, Los Angeles. Her dissertation examines the history of leprosy in Korea. Her interests mainly concern history of international and global health and its relations to Korea and Northeast Asia. She is currently completing her manuscript "Leprosy in Korea: A Global History."

Sonja M. Kim is an associate professor of Asian and Asian American Studies at Binghamton University of the State University of New York. She is the author of *Imperatives of Care: Women and Medicine in Colonial Korea* (University of Hawai'i Press, 2019). Her current projects explore professional ethics in medicine and nursing, health investigations of the urban poor, and public health regimes in Cold War Koreas.

Robert Ji-Song Ku is an associate professor of Asian and Asian American Studies at Binghamton University of the State University of New York. He is coeditor of *Pop Empires: Transnational and Diasporic Flows of India and Korea* (University of Hawai'i Press, 2019) that juxtaposes the popular culture regimes of Korea, India, and the United States. He is also the author of *Dubious Gastronomy: Eating Asian in the USA* (University of Hawai'i Press, 2014) and coeditor of *Eating Asian America: A Food Studies Reader* (2013). He coedits the Food in Asia and the Pacific series for the University of Hawai'i Press.

Jieun Lee is a cultural anthropologist working at the intersection of medical anthropology and Science and Technology Studies (STS). Her research focuses on the relation between technoscientific knowledge production, ordinary practices and affects in and outside of labs, and changing conceptions of body, time, and life in contemporary South Korea. She received her PhD in anthropology at the

University of California, Davis. Her dissertation, "The Promises of Biology and the Biology of Promises: An Ethnography of the Korean Stem Cell Enterprise," attends to how promises of better futures are anchored in the biological notion of potential in the labs, the offshore stem cell treatment markets, anti-aging business, and the stock market. Currently, she is an assistant professor in the Department of Cultural Anthropology at Yonsei University.

Hye-ri Oh is a research professor in the Research Institute for the Visual Language of Korea at Myongji University in Seoul, South Korea. Trained as a photographer and modern and contemporary art historian, she earned her PhD from Binghamton University of the State University of New York in 2014. Her research interests include the histories of art, photography, and visual culture in East Asia. She is working on a book manuscript that examines the emergence and transformation of the conception of photography in Korea and Japan from the late nineteenth century through the era of Japanese colonialism. Her article "Translating Photography: The Migration of the Concept of Sajin from Portraiture to Photography" was published in the journal *History of Photography*.

Haerin Shin earned her PhD in comparative literature at Stanford University, working on contemporary American, Korean, and Japanese literature, culture, critical theory, and other forms of media such as film, animation, and graphic narratives. Her research focuses on ontology and technology, cognitive literary theory, psychoanalytic criticism, cyborg theories, the posthuman, speculative fiction, and Asian American literature. She is currently completing a book titled *Dialectic of Spectrality: A Transpacific Study on Being in the Age of Cyberculture, 1945–2012*, in which she notes that the advent of computers, the internet, and networked mobile devices throughout the latter half of the twentieth century has brought abstracted flows of data to the fore of social interaction and communication. She is also in the process of designing her next book, which will address immortality and consciousness.

Theodore Jun Yoo is a professor in the department of Korean Language and Literature at Yonsei University in Seoul, South Korea. He is the author of *The Politics of Gender in Colonial Korea* (2008), *It's Madness: The Politics of Mental Health in Colonial Korea* (2016), and *The Koreas: The Birth of Two Nations Divided* (2020).

Index